PACKAGING DESIGN

PACKAGING DESIGN

AN INTRODUCTION

Lászlo Roth

VAN NOSTRAND REINHOLD
New York

Library of Congress Catalog Card Number 89-5329
ISBN 0-442-31863-4

Printed in the United States of America

Van Nostrand Reinhold
115 Fifth Avenue
New York, New York 10003

Van Nostrand Reinhold International Company Limited
11 New Fetter Lane
London EC4P 4EE, England

Van Nostrand Reinhold
480 La Trobe Street
Melbourne, Victoria 3000, Australia

Nelson Canada
1120 Birchmount Road
Scarborough, Ontario M1K 5G4, Canada

16 15 14 13 12 11 10 9 8 7 6 5 4 3 2 1

Library of Congress Cataloging-in-Publication Data

Roth, Lászlo.
Packaging design : an introduction / Lászlo Roth.
p. cm.
Bibliography: p.
Includes index.
ISBN 0-442-31863-4
1. Packaging—Design. I. Title.
TS195.4.R68 1989
688.8—dc19 89-5329
CIP

To my packaging design students at F.I.T.
and in memory of my student Regina Siepmann 1963–1988

CONTENTS

PREFACE

Ours is a technological culture characterized by ever-decreasing handwork. The majority of workers are employed in industries that have become so specialized that a worker may devote his or her entire career to a single operation or action. In contrast, design and craft give individuals an opportunity to feel fulfilled. The designer must conceive an idea and then carry it through every stage to completion, using a multitude of tools and materials. It is in this complete involvement which creative activity offers that the student comes to know him- or herself fully and realizes creative potential.

Design education, unfortunately, is often handicapped by "identity" problems. That is, the way in which professionals describe themselves—graphic designers, industrial designers, environmental designers, architects—becomes more important than the work they actually do. Yet the study of design in all its aspects offers a broad spectrum of opportunities that can be gratifying for the student. And the study of packaging design is especially fascinating.

Young designers often have limited experience and training in packaging design and technology. Students and even practicing designers have a tendency to cling to familiar topics and skills. But skills of all kinds are essential to creative expression. Design capability results from a fusion of skills, technologies, and materials. The student can only benefit from a detailed knowledge of the skills and technologies of packaging design.

This basic guide to packaging describes everything you need to know about this exciting industry. It is an introduction to the world of packaging design and technology. It has been written by a designer for designers in the sincere hope that it may point the way to a challenging and meaningful career.

ACKNOWLEDGMENTS

My sincere thanks go to the following people who helped me in the course of preparing this book:

My packaging design students and members of the Package Design Council International

Fred Roesler, Tenneco Polymers, Inc.

George W. McIntyre, Reed-Prentice, Div. Package Machinery Co.

Rita J. Simpson, Brick-Pac, Inc.

Philip Wingard, Bekum Plastic Machinery Co., Inc.

Mark B. Ponsky, Hoover Universal, Inc.

James E. Bolls, Autoroll-Dennison Corp.

Lauren R. Paulsen, *Packaging Digest*

O. A. Freshour, Ludlow Flexible Packaging

James DiMarco, Packaging Systems Corporation

Lee Reinhart, Hoan Products Ltd.

Massimo Vignelli, Vignelli Designs

William J. Vituli, The Great Atlantic & Pacific Tea Co., Inc.

Frank Csoka, Designer

Ivan Biro, Sculptor

George Wybenga, Signage

Thomas Giaccone, Designer

Elizabeth Downey, Designer

Gabrielle Roth, Research

Miles David Sebold, Photographer

Jane Chope, Landor Associates

Annette Green, The Fragrance Foundation

Leslie Weller, Elizabeth Arden

Denise Ortel, Helena Rubinstein, Inc.

Richard Rabkin, Ideal Toy Corporation

Steve Doyal, Hallmark Cards, Inc.

Alice Simpson, Designer

Joan Gillman Negrin, Designer

Jack Schecterson, Designer

Anita Mott, Designer

1

INTRODUCTION

(Photo courtesy of the Great Atlantic & Pacific Tea Company, Inc.)

Packaging in Modern Society / Packaging and Marketing / The Designer's Role

PACKAGING has always been an important part of a product and thus an important industry. Figure 1-1 shows some creative and appealing packages of nineteenth-century products. But since World War II, the packaging industry has become a significant economic force in the advanced industrial nations. For one thing, the manufacture, use, and disposal of packaging accounts for a large

Figure 1-1. Early American packages (nineteenth century).

proportion of the activities of modern businesses. For another, numerous business functions, including marketing, advertising, and the production of point-of-purchase and promotional materials, are dependent on the packaging industry.

Consider these facts related to the U.S. economy and the packaging industry: In the United States alone, more than 1 million people are employed in some aspect of packaging—more than in any other single industry. These people work in the more than 300,000 companies in 200 manufacturing industries that perform packaging functions. And they use materials and equipment provided by more than 5000 suppliers. Significantly, corporations spend more than $50 billion a year on packaging. In fact, 75 percent of all goods purchased by consumers in the United States are distributed in packages.

Packages perform protective and distributive functions; and it is these functions that help keep the cost of finished goods down while costs of manufacturing products are constantly escalating.

PACKAGING IN MODERN SOCIETY

Changes in key characteristics of the American lifestyle and population have had dramatic effects on packaging. Among these are demographic changes, such as the increased median age of the population, increased disposable income, increased life expectancy, a greater amount of leisure time, increased employment of women, and a shift from urban to suburban residence. In addition, new technologies, the role of government regulation, ecological concerns, and consumerism affect packaging.

But perhaps the most important influence on packaging is the use of the automobile for shopping. It is the automobile that has led to the establishment of the great shopping centers, with their multiple stores and services, which are at the heart of American buying patterns today.

Another key influence on packaging is the expansion of supermarkets and retailing chains throughout the country. These enterprises have demanded a new look at the methods of selling, packaging, and new-product introduction (Fig. 1-2). Supermarkets, for example, offer up to 10,000 items, of which 20 percent are *loss leaders,* that is, sold at or below cost to bring shoppers into a store.

Figure 1-2. Drug store product display within a supermarket. *(Photo courtesy of the Great Atlantic & Pacific Tea Company, Inc.)*

Once in the store, shoppers will be enticed to buy up to 50 percent more on impulse. Special offerings, such as premiums and samples, encourage impulse buying. But designers can also influence such buying with strong, exciting graphic design. Packaging clearly plays a central role in generating the impulse to buy.

PACKAGING AND MARKETING

Today, package design and product marketing are precise sciences, with the designer being an integral part of the marketing function. The competent package designer knows that creative marketing is based on anticipating future trends. Knowing the buying habits of consumers can be a great help in planning the graphic and structural design of a package or point-of-purchase (POP) display.

Point-of-purchase displays are specifically created to attract the consumer's attention at the point of sale and are among the most effective means of selling products (Fig. 1-3). Packages therefore need to be designed so that they can be used effectively in such displays.

The package design must also be tied in effectively with newspaper, magazine, and television advertising. Ads that effectively depict the package enable consumers to identify the product through the package. *Logos,* that is, the graphic depiction of company or brand names, also play an important role in identification and recognition.

Figure 1-3. Point-of-purchase display. *(Photo courtesy of Hallmark Cards, Inc.)*

THE DESIGNER'S ROLE

In view of the importance of packaging, it is safe to say that the package designer occupies a central role in modern manufacturing. Yet few people understand the role of the package designer. The designer must do more than simply enclose a product in a container. He or she must create a package that is unique, is aesthetically pleasing, helps sell the product, runs on existing packaging machinery, and meets other needs of the client. As if this were not enough, the designer must satisfy five "masters":

1. Makers of the product—that is, product designer—with logo for corporate identity, consumer product manufacturing corporations (cosmetics, food, etc.), equipment for package filling and production

2. Makers of the package—that is, package designer—with graphic and structural design, package production, which includes the printing and molding processes

3. Retailers and intermediaries, with speciality designs in supermarkets, department stores, boutiques, and so on

4. Consumers, with marketing surveys and demographic studies

5. Regulatory agencies such as the Food and Drug Administration (FDA) and the Federal Trade Commission (FTC), with knowledge of correct regulations

Only a designer with a broad background in design and considerable technical knowledge can meet this challenge.

Years ago packaging was a purchasing function, and package designers were regarded as commercial artists with no expertise beyond graphic design. In most cases they were called in to make the package "look good" after key marketing decisions had already been made. Today, in contrast, the designer is viewed as a marketing professional who does not just come up with a design but who also creates solutions to marketing problems.

A major reason for this change in how designers are seen is the recognition that packaging is the least expensive form of advertising. Every package is, in effect, a five-second commercial (Fig. 1-4).

Figure 1-4. Cat box. J. C. Chou, designer.

A well-planned packaging program requires a team of experts. Graphic and product (industrial) designers, marketing managers, salespeople, and manufacturing and distribution experts must participate. The designer must integrate the needs and requirements of all these experts in creating a suitable package.

Too many designers fail to recognize the needs and wants of a changed society. Today, the designer must be tuned in to demographic changes, new technologies, the role of government regulation, ecological concerns, and consumerism. Unless serious attention is given to developments in these and related areas, the designer will be unable to create a truly successful package.

2

THE MAKING OF A DESIGNER: IDEAS AND TECHNIQUES

The Designer's Qualifications / Design as an Aspect of Marketing / Ideas and Concepts / Tools of the Designer / Typography / Color / Photography / Phases of a Design Project / Computers and the Package Designer / Preparing Your Portfolio

THE work of the package designer begins with an idea, or visual image. But where does the visual image come from? How is it transformed from a rough sketch into a three-dimensional object? These questions form the basis of creativity, the fundamental asset of the package designer, and as such deserve careful study.

To appreciate the importance of creativity, consider Leonardo da Vinci, one of the most inspired creative thinkers of all time (Fig. 2-1). Da Vinci is well known for his art, but art received only part of his attention. The full scope of da Vinci's creative activity was much broader. Leonardo da Vinci observed life from many perspectives. He was a scientist, an inventor, and a painter.

It was generally agreed among his contemporaries that whenever da Vinci undertook a creative task he would produce something unusual, something wonderful to behold. And he did. His notebooks contain 5000 pages of extraordinary artistic and scientific research and inquiry. They include sketches of scientific instruments, hydraulic engines, sections of

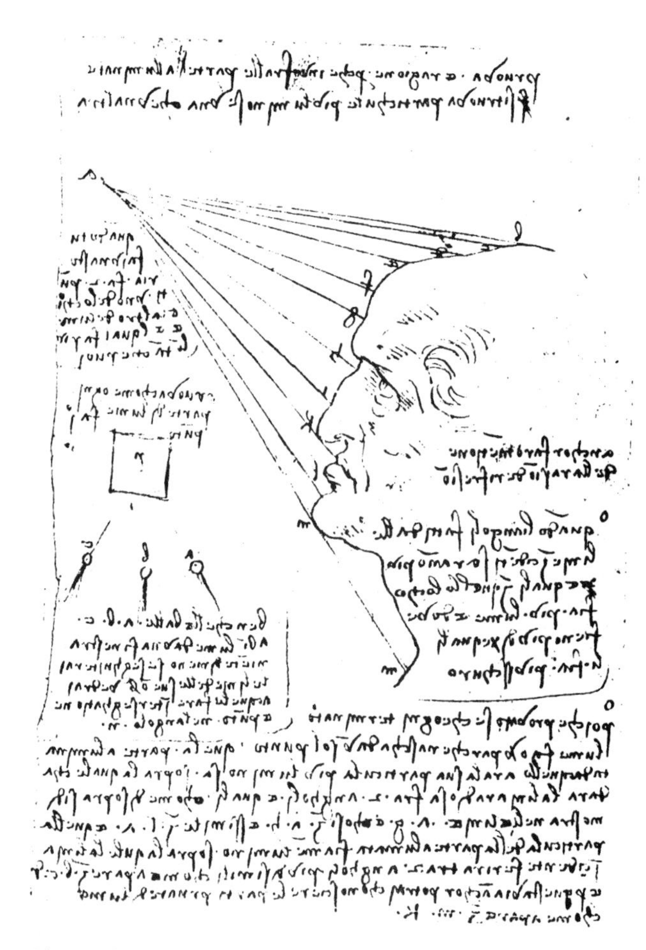

Figure 2-1. From the notebooks of Leonardo da Vinci. *(Reproduction from the author's collection.)*

Figure 2-2. From the notebooks of Leonardo da Vinci. *(Reproduction from the author's collection.)*

bones, muscles, leaves, trees, and cloud formations. An example is shown in Figure 2-2.

The more da Vinci studied, the deeper his wisdom and the sharper his skills became. Each new undertaking implied a new and unique design. Although he didn't work by a formula, his creativity was based on careful, systematic observations and studies. In this respect da Vinci was not unlike the skilled designers of today, as will become clear in the following discussion.

THE DESIGNER'S QUALIFICATIONS

The chief qualification of a good designer is the ability to communicate the concept to the client. To do so, a skilled de-

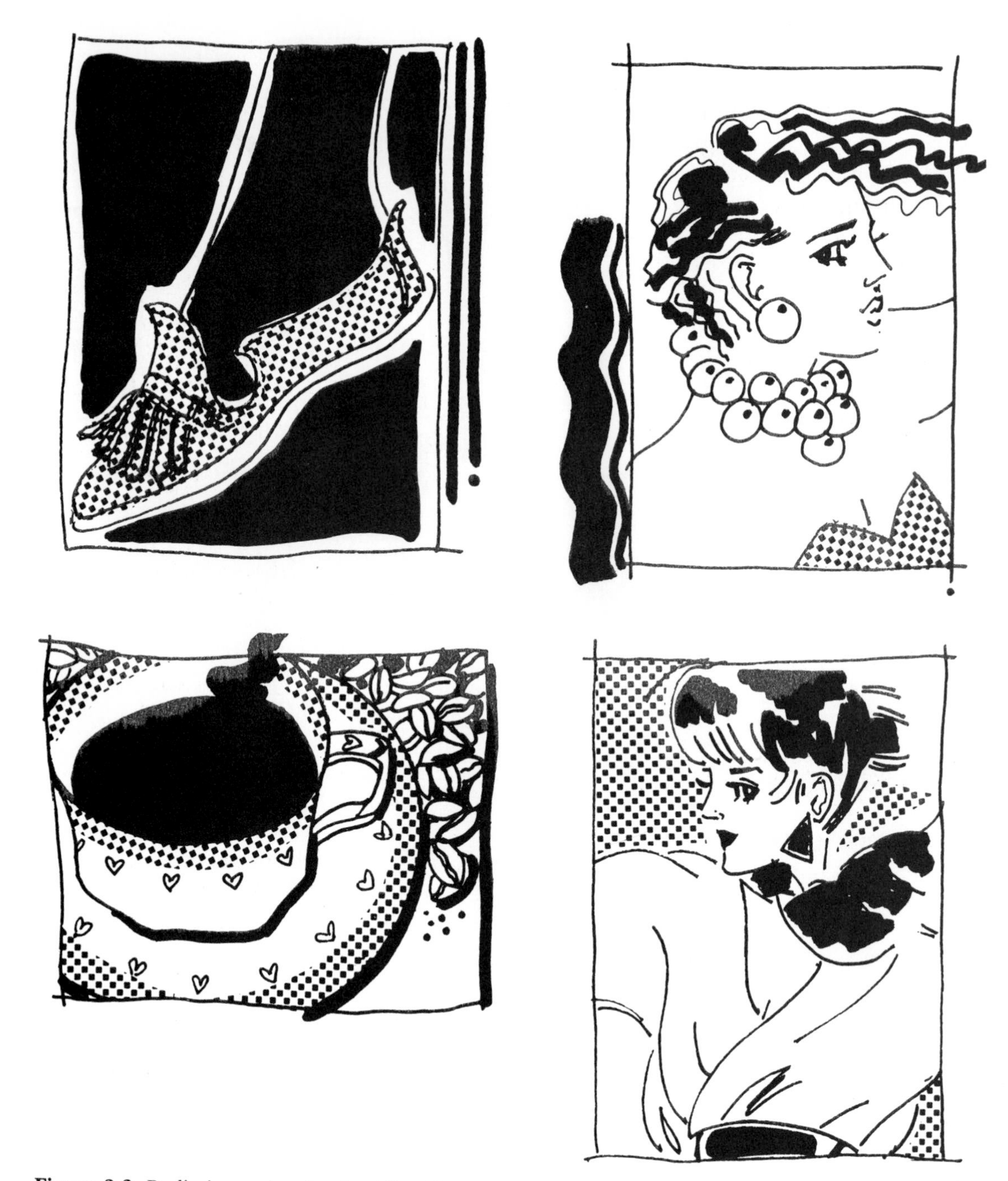

Figure 2-3. Preliminary sketches by Alice Simpson.

signer—such as Leonardo da Vinci—must have varied knowledge. He or she must be familiar with the history of art, especially artistic styles. A designer must be a reader, observer, and critic of the art scene, including theater, cinema, music, and dance. A designer should visit galleries and museums. In short, a designer must be continually tuned in to potential sources of ideas.

A key task of the designer is producing visual concepts that can be understood by people who are not visually oriented but whose judgment and approval are required if the project is to be a success (Fig. 2-3). Therefore, it goes without saying that a designer should be able to draw.

Figure 2-4. Candy box. *(Photo by Push Pin Studio Group. Seymour Chwast, designer.)*

Some of the finest contemporary designers are illustrators as well as gifted designers (Fig. 2-4). There is a saying in the design field that one can make a designer out of an illustrator, but it is difficult to make an illustrator out of a designer. Again, sound training in the art of drawing is the first step to becoming a successful designer. In recent years, major art schools have added more hours of drawings to their graphic design curricula. The aspiring designer should take as many such courses as possible.

Knowledge is the foundation of drawing. Artists draw what they know, not (as commonly suggested) what they see. An artist is trained to see with the mind; the eyes are merely tools. While the painter usually works from live models, the designer must work without a model. This fact increases the importance of knowing the basic structure of objects, landscapes, and human and animal figures (anatomy). The artist must develop a perception that recognizes shapes and masses, not just lines, and must learn the art of simplifying, that is, of distinguishing between what is important and what is incidental and irrelevant.

Finally, since the designer's concepts are produced in their final form by a professional illustrator or photographer, the designer must understand the techniques of illustration and photography. Otherwise, there are likely to be costly, time-consuming delays.

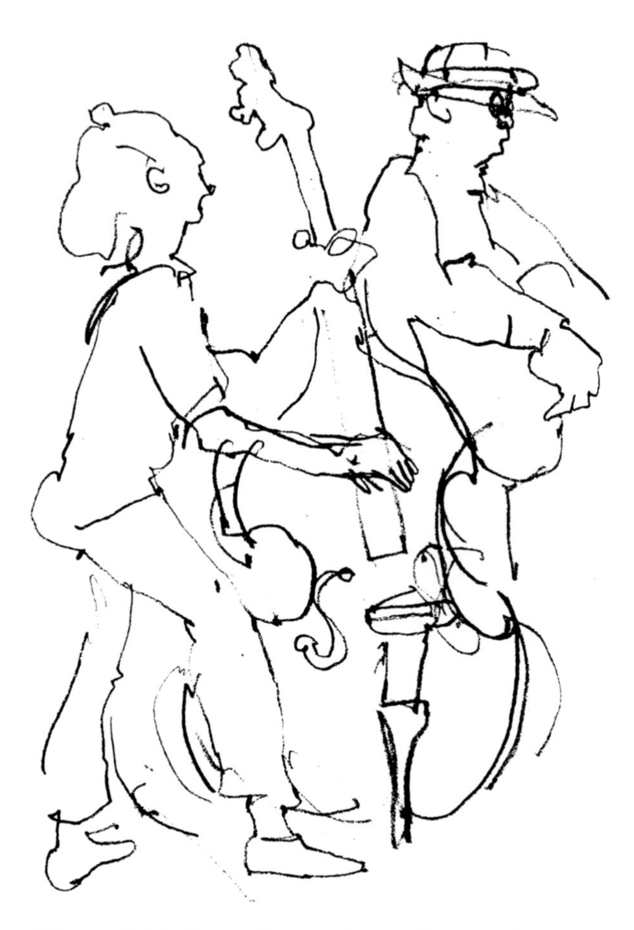

Figure 2-5A. From the notebook of Joan Gillman Negrin.

Basic Courses for Design Students

Here is a list of suggested basic courses for students of design. Note that they provide the skills and expertise needed to be a good designer.

1. A thorough and extensive drawing course (life drawing) that meets at least three hours a week. When taking the course, make sure you have a sketchbook with you at all times (Fig. 2-5A). Draw wherever and whenever you can, everything and anything. Never tear out pages of "bad" work; keep all of your sketches to see how you are progressing. Use any medium you prefer—pencils, markers, pens, and so on. Your notebooks, like da Vinci's, are a record of your personal observations and studies (Fig. 2-5B).

2. Learn to work in various media, including tempera, oil, acrylic, pastels, watercolors, and above all markers. After you have learned how to draw, take a course on rendering with markers. Marker rendering requires special techniques that depend on good drafting skills. (See the section on Tools of the Designer in this chapter.)

3. Photography is a must for the designer. Buy a good 35 mm camera and a projector, and take a basic technical course in photography. Both the knowledge and the equipment will be important assets in your work.

4. A talent for lettering can get you great jobs. Brush up on your lettering skills and learn comp lettering.

5. Study art history; develop your taste by learning about great art of the past and present. But also maintain a strong and active interest in new ideas.

Figure 2-5B. From the notebook of the author.

Ordinary Versus Good Designers

Some designers—who tend to be imitative and extremely verbal—are in business simply to make money. Others feel that design is a form of personal artistic expression, usually carried out at the client's expense. Then there is the "Yes sir, we can do it, no problem" type; these designers will do exactly what the client dictates and are likely to come up with cliché solutions.

In summary, a good designer is a craftsperson, an illustrator, a marketer, a photographer, a letterer, and—above all—a verbal as well as visual communicator.

DESIGN AS AN ASPECT OF MARKETING

In addition to possessing the basic qualifications of a good designer, a package designer must have skills specific to designing for commerce and industry. Designing for commerce and industry is part of the process of selling products, that is of marketing. Before beginning the actual project, the designer therefore has to do extensive preliminary research work.

Know Marketing Procedures and Techniques

Since packaging is a marketing function, the designer needs to be familiar with marketing procedures and techniques. Marketing can be thought of as a set of procedures for developing products and planning sales strategies and campaigns. Indeed, the first thing a newly hired designer should do is study the market by visiting retail stores and interviewing managers, sales personnel, and if possible consumers. He or she should also visit the firm's marketing department and meet the group that will help him or her develop ideas into concepts.

Know the Product

For a product to sell, the designer must know it intimately. He or she must be thoroughly familiar with the needs, preferences, tastes, purchasing power, and buying habits of the consumer. The designer also must be aware of the client's needs and problems. Marketing problems, competitive pressures, and the client's budget must all be considered in planning a design project.

Be Aware of Competitive Products

The information obtained from managers, sales personnel, consumers, and the marketing staff must be evaluated from a competitive standpoint. The designer must always remember that a package is never alone. It is surrounded by other packages, usually those of competing products. It is important to compare these competing packages with the client's existing packages (Fig. 2-6).

Consider the Message

When formulating the design strategy, the designer must give attention to the message or messages the package is to deliver to potential consumers. Packages can deliver messages about the product, the brand, the product category, the typical customer, or the benefits offered by the product. The package can also project uniqueness, create an image, or increase the impact of other promotional tools. The package can also send subliminal messages. For example, color, shape, size, and texture may suggest luxury (em-

Figure 2-6. The designer must be aware of competition. The pretty package in the store is never alone. *(Photo courtesy of the Great Atlantic & Pacific Tea Company, Inc.)*

bossing, foil, or unusual paper stock). Transparent packages (visible merchandise), structural design (unusual shapes), or reusable packages (bottles, jars, or boxes) also send different messages. Glass containers, despite such drawbacks as fragility and excess weight, suggest that the product is of superior quality.

The package is the symbol of a total marketing effort; it is the visual and physical evidence of the product being sold. Therefore, it is important to keep in mind that in addition to just selling the product packaging has a lot to do with brand loyalty. Put a defective valve on an aerosol can and you may lose a loyal customer.

Keep Well Informed

The designer also has homework to do to keep up with industry developments. He or she should read the trade publications and books and talk to packaging suppliers, who are always well informed on the latest techniques and materials.

IDEAS AND CONCEPTS

On the basis of all the preparatory work, the designer may produce a promotional device, a package, or a product. The core of these efforts is the idea. The visual aspect of an idea is a concept.

An idea and its concept can come from anywhere or anyone. But an idea for a specific purpose can only come from a trained, disciplined mind, which is accustomed to thinking realistically and to rationalizing the visual aspect of the idea. *Rationalizing the visual aspect* means creating an effect (with color or shape), a reason for a particular package (e.g., using squeeze tubes instead of jars for tomato paste), or a need for a product (such as impulse items).

In sum, package designers must be equally at home with the aesthetic, technical, and marketing aspects of a project. The designer is an artist who can begin with an idea and draw, photograph, model, or build a prototype in order to present a concept that the client can understand and evaluate.

TOOLS OF THE DESIGNER

In order for designers to do their work properly, they need the right tools. The following is a list of basic supplies recommended by most art schools:

1 Red fiber expanding envelope 17″ × 22″, or a portfolio

1 Tracing pad 14″ × 17″ (not vellum)

1 Visualizer pad 14″ × 17″

1 Bristol pad 14″ × 17″, 2 or 3 ply

1 Steel ruler 24″

1 Transparent plastic triangle 6″, 45° angle

1 Transparent plastic triangle 12″, 30°, 60°, 90° angles

1 Sandpaper block

1 Kneaded eraser

1 Roll masking tape (¾″, ½″, or 1″)

1 Pint rubber cement in plastic jar

1 Pint rubber cement thinner

1 Squirt can for thinner

1 Rubber cement pick-up

1 Compass, Alvin 805V or V203

1 Ruling pen, Kern #3003, or Charvos #11-0052

1 Package compass blades (flat blade)

1 Pair scissors, Fiskars 5″

1 Bottle india ink black/white proof, Pelican, Higgins, Artone

1 Crowquill pen holder and pen nibs

3 Graphite pencils (or lead holders) HB, 2H, 4H

1 X-Acto knife, #11 blades

1 Replaceable blade mat knife

1 Plastic bottle white glue (Sobo, Elmer's)

6 Single edge razor blades

1 Set of cool gray designer markers

Assorted color markers (your instructor's list)

2 Black Pilot razor point pens

1 Circle cutter with blades

1 Hole puncher

Assorted sable brushes (your instructor's list)

Assorted designer's colors (your instructor's list)

Markers

One tool deserves special mention—the marker pen. The marker pen has been one of the most important recent developments in graphic media. In the 1940s the refillable felt point pen was developed. In the early 1950s the first color markers appeared and were an instant success with artists and designers. In fact, this simple tool created a veritable revolution in the graphic arts.

Markers are formulated with dyes, not pigments. They come under different brand names, some in as many as 300 transparent colors. They may be water-, thinner-, or acrylic-based. Their tips range from extremely fine to very wide chisel points.

Markers are amazingly versatile. A trained designer can achieve effects re-

Figure 2-7. Marker sketch by Thomas Giaccone, designer.

Figure 2-8. Marker rendering of a printing toy. *(Photo courtesy of Ideal Toy Corporation.)*

markably similar to those obtained with pen and ink, wash, watercolor, pastel, and even oils (Figs. 2-7 and 2-8). So important has this medium become that marker rendering is now a separate course in art schools.

Necessity for Practice

Even the best tools and media will not make a designer or artist out of someone who is unwilling to learn by doing. The only form of practical art education is experience. This means practice in sketching, drawing, and painting, together with exposure to all the arts. The dream of every creative designer is to develop a style that is recognized, appreciated, admired, and imitated. Again, the secret of developing a distinctive style is *practice*!

The Five Don'ts and Dos

To help you practice, here are five things not to do as well as what you should do:

1. *Don't* work with short, fumbling lines. *Do* learn to work with spontaneous, continuous lines. And *do* use soft pencils (HB, #2) or markers, not pen and ink.

2. *Don't* try to show too much. *Do* show close-ups of heads and shoulders rather than full figures. *Do* suggested backgrounds rather than put in a lot of confusing details.

3. *Don't* try to indicate perspective, except in a simple form.

4. *Don't* get carried away with the drawing. *Do* draw only to illustrate the idea.

5. *Don't* copy. There is no surer way to shut off individual creativity and learning. *Do,* however, use or adapt pictorial reference material. In fact, it's a good idea to build your own picture reference files (sometimes called "swipe files"); these can save you a lot of valuable research time. Your files should be arranged by subject (e.g., animals, architecture, humor, decorative art). The "swipes" in your file should inspire you, teach you, and help you develop you own style.

TYPOGRAPHY

Once the designer has done the preliminary market research, has an idea of what the package is to look like, and is comfortable with design tools, he or she can begin to consider the visual aspect of the idea, that is, the concept. The first thing to think about is typography.

Typography is the style, arrangement, or appearance of typeset matter. It determines legibility and the ease with which we recognize letters and numbers. For the designer, typography involves the selection of typefaces for specific purposes, usually with a specific aesthetic goal. It entails, for example, using correct, pleasing type to create logotypes for corporate identities, using easy-to-read type, choosing strong type for headlines and product (brand) names, and creating new typefaces for specific purposes. Since typography appeals to the eye, the principles of design also apply to typography.

Historical Background

From an historical standpoint, typography is the study of the written word as well as the printed word. The earliest ancestors of our so-called Roman type are the stone-cut inscription on Trajan's column (114 A.D.) and the hand-lettered manuscripts of Italy and Northern Europe (fourth to eighth centuries). The first typographers, in fact, imitated and adapted these manuscripts. Through the centuries, variations of the most legible and beautiful scripts—uncial and half uncial, caroline minuscule (ninth and tenth centuries), and the humanistic book hand (sixteenth century)—evolved into the forms on which present typefaces are based.

Typefaces

There are about 6000 typefaces, enough to meet virtually any typesetting need. But it is impossible to choose and specify type intelligently unless one knows many of them by sight. To begin to learn them, the designer can study typefaces in the light of the history and influences that gave them their shape and character. Following this approach, we can classify typefaces into five groups: text or black letter, roman, italic, cursive and script, and block letters. Figure 2-9 gives examples of each of these typefaces.

Text or Black Letter

Text is the type that was used by Gutenberg and his contemporaries. It developed out of the fifteenth-century scripts (Fig. 2-10). Text is sometimes called Old English because it was used by William Caxton, the first English printer.

Black letter is also known as Gothic. Gothic typefaces consist of about 60 families that were developed over many years. There are both regular Gothics and square serif Gothics (see your type books).

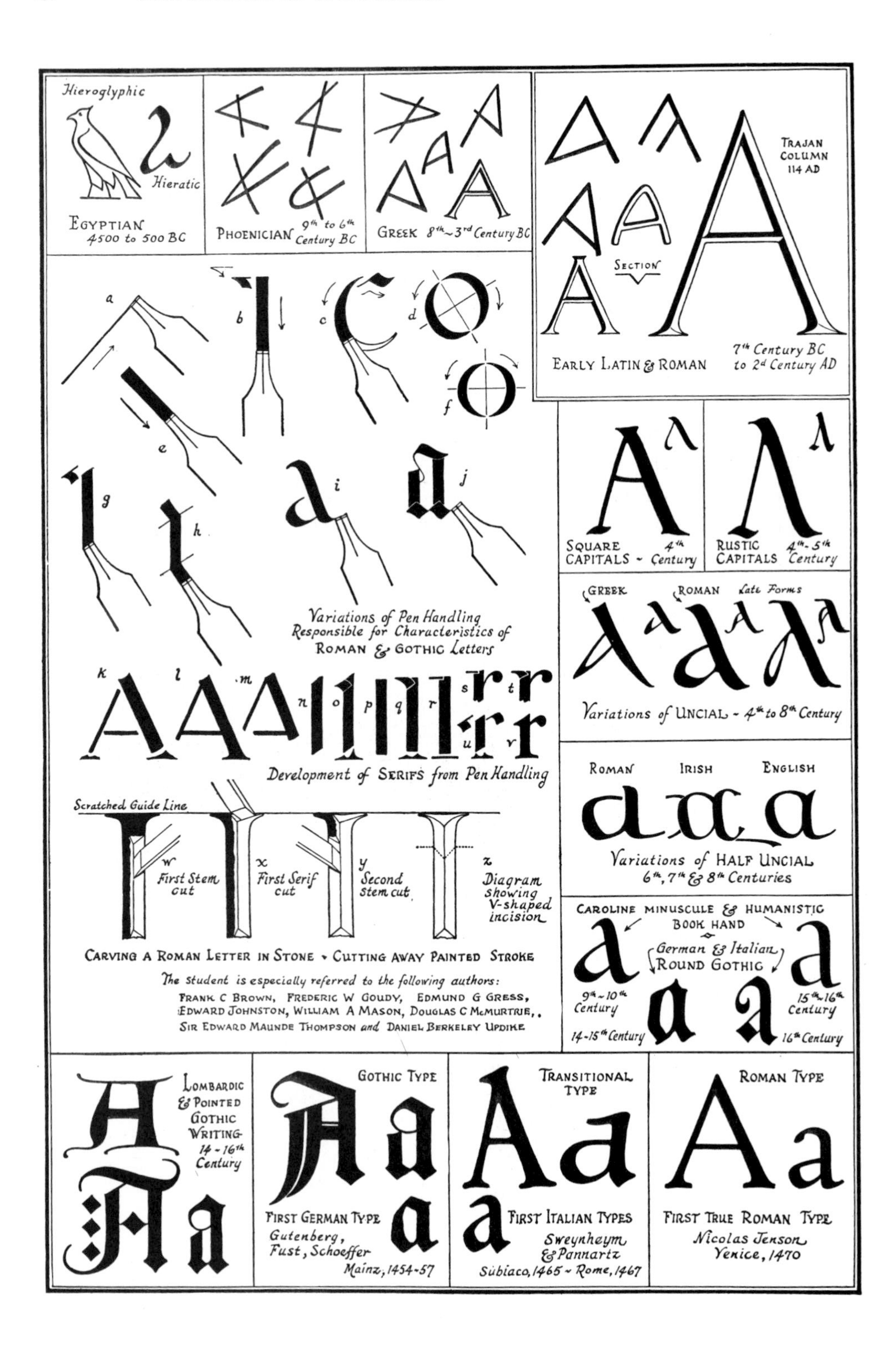
Hieroglyphic
Hieratic
EGYPTIAN 4500 to 500 BC
PHOENICIAN 9th to 6th Century BC
GREEK 8th–3rd Century BC
TRAJAN COLUMN 114 AD
SECTION
EARLY LATIN & ROMAN 7th Century BC to 2d Century AD
Variations of Pen Handling Responsible for Characteristics of ROMAN & GOTHIC Letters
Development of SERIFS from Pen Handling
SQUARE CAPITALS – 4th Century
RUSTIC CAPITALS 4th–5th Century
GREEK ROMAN Late Forms
Variations of UNCIAL – 4th to 8th Century
ROMAN IRISH ENGLISH
Variations of HALF UNCIAL 6th, 7th & 8th Centuries
Scratched Guide Line
First Stem cut
First Serif cut
Second Stem cut
Diagram showing V-shaped incision
CARVING A ROMAN LETTER IN STONE • CUTTING AWAY PAINTED STROKE
The student is especially referred to the following authors:
FRANK C BROWN, FREDERIC W GOUDY, EDMUND G GRESS,
EDWARD JOHNSTON, WILLIAM A MASON, DOUGLAS C MCMURTRIE,
SIR EDWARD MAUNDE THOMPSON and DANIEL BERKELEY UPDIKE
CAROLINE MINUSCULE & HUMANISTIC BOOK HAND
German & Italian ROUND GOTHIC
9th–10th Century
14–15th Century
15th–16th Century
16th Century
LOMBARDIC & POINTED GOTHIC WRITING 14–16th Century
GOTHIC TYPE
FIRST GERMAN TYPE Gutenberg, Fust, Schoeffer Mainz, 1454–57
TRANSITIONAL TYPE
FIRST ITALIAN TYPES Sweynheym & Pannartz Subiaco, 1465 – Rome, 1467
ROMAN TYPE
FIRST TRUE ROMAN TYPE Nicolas Jenson Venice, 1470

Figure 2-9 (opposite page). Condensed outline of root forms of the Roman alphabet based on a single character. *(From Thomas Blane Stanley,* The Technique of Advertising Production, *p. 149. © 1954 Prentice Hall, Inc., Englewood Cliffs, N.J. Reprinted with permission.)*

leuauerunt anforã inter terram et celũ.
Et dixi ad angelũ qui loquebat̃ ĩ me.
Quo iste deferũt anforã? Et dixit ad
me. Ut edificet̃ ei dom⁹ ĩ terra sennaar:
et stabiliat̃ et ponat̃ ibi sup basem suã.
Et cõuersus sum: et leuaui
oculos meos et vidi: et ecce qua-
tuor quadrige egredientes de medio du-
orũ montiũ: ⁊ mõtes montes erei. In
quadriga prima equi rufi: in quadriga scđa equi nigri in quadriga tercia
equi albi. ⁊ in q̃driga quarta equi varij ⁊ fortes. Et respõdi: ⁊ dixi ad angelũ qui loquebat̃ in me. Quid sunt hec
dñe mi? Et respondit angelus: et ait
ad me. Iste sunt q̃tuor venti celi: qui e-
grediũtur ut stent corã dñatore omĩs
terre. In qua erãt equi nigri egrediebã-
tur in terrã aquilonis: et albi egressi
sunt post eos: et varij egressi sunt ad
terrã austri. Qui aũt erãt robustissimi
exierũt: ⁊ querebãt ire ⁊ discurrere per o-
mnẽ terrã. Et dixit. Ite: ⁊ perambulate
terrã. Et perambulauerũt terrã. Et vo-
cauit me: ⁊ locut⁹ est ad me dicẽs. Ecce qui egrediũtur in terrã aquilonis:
requiescere fecerũt spiritũ meũ in terra
aquilonis. Et factũ est verbũ dñi ad
me dicens. Sume a transmigratione
ab oldai et a thobia et ab idaia et ve-
nies tu in die illa et intrabis domũ io-
sye filij sophonie qui venerũt de babi-
lone: ⁊ sumes aurũ ⁊ argentũ ⁊ facies
coronas et pones in capite ihesu filij
iosedech sacerdotis magni: ⁊ loquens
ad eũ dicens. Hec ait dñs exercituum
dicẽs. Ecce vir: oriens nomẽ eius. Et
subter eũ orietur: et edificabit templũ
dño: et ipe extruet templũ dño. Et ipse
portabit gloriã: ⁊ sedebit ⁊ dñabitur
sup solio suo. Et erit sacerdos sup solio suo: et cõsiliũ pacis erit inter illos
duos. Et corone erũt helem et thobie
et idaie ⁊ hen filio sophonie: memori-
ale in templo dñi. Et qui procul sunt
venient et edificabũt in templo dñi: et
scietis quia dominus exercituum misit me ad vos. Erit aũt hoc: si auditu
audieritis vocẽ dñi dei vestri.
Et factũ est in anno quarto darij
regis: factum est verbum dñi ad
zachariã in quarta mensis noni qui
est casleu. Et miserũt ad domũ dei sa-
rasar et rogomelech et viri qui erãt cũ
eis ad deprecandã faciẽ dñi: ut dicerẽt
sacerdotib3 dom⁹ dñi exercituũ ⁊ ꝓphe-
tis loquẽtes. Nũquid flendũ ẽ michi
in quinto mense: vel sanctificare me
debeo sicut feci iam multis ãnis? Et
factũ est verbũ dñi exercituũ ad me dicens. Loquere ad omnẽ ꝓplm terre: et
ad sacerdotes dicens. Cũ ieiunaretis
et plangeretis in quinto ⁊ septimo per
hos septuaginta ãnos: nũquid ieiu-
niũ ieiunastis michi? Et cum comedi-
stis ⁊ bibistis: nunq̃d nõ vobis come-
distis et vobismetipsis bibistis? Nũquid nõ hec sunt verba que locutus
est dñs in manu ꝓphetaꝝ prioꝝ cũ ad-
huc iherusalẽ habitaret̃ et esset opulẽ-
ta: ipa ⁊ urbes in circuitu ei⁹: et ad au-
strũ ⁊ in cãpestrib3 habitaretur? Et factũ est verbum dñi ad zachariã dicẽs.
Hec ait dñs exercituũ dicens. Iudiciũ
verũ iudicate: et misericordiã et misera-
tiones facite unusquisq3 cũ fratre suo.
Et viduã et pupillũ ⁊ aduenã et pau-
perem nolite calũniari: ⁊ malũ vir fra-
tri suo non cogitet in corde suo. Et no-
luerũt attendere et auerterunt scapulã
recedẽtes: ⁊ aures suas aggrauauerũt
ne audirent: et cor suum posuerũt ut
adamãtem ne audirẽt legẽ et verba q̃
misit dominus exercituũ ĩ spiritu suo

Figure 2-10. A page from Gutenberg's original Bible, 1452.

Roman

Roman type is also derived from the fifteenth-century scripts. Two German typographers, John and Wendelin of Spire, and a Frenchman, Nicholas Jensen, all of whom were working in Venice around 1470, were the originators of this truly beautiful type.

Italic

The type that we know as italic was first used about 1500 by the famous Italian printer Aldus Manutius. At first, lowercase italic letters were used with small Roman uppercase letters. The slanted, or italic, capital did not come into general use until about 1538, more than 20 years after Aldus' death. (Uppercase and lowercase letters are discussed later in this section.)

Cursive and Script

The cursive and script styles are derived from informal running handwriting. Both of these types display a tendency toward inked letters and flowing strokes.

Block Types

The category of block types includes a large group of typefaces, mostly without serifs and approximately the same weight throughout.

Block characters are actually simplified, highly legible Roman types that were introduced in England about 1800. In our own time, a remarkably elegant type based on block types, Futura, has been developed. Futura, which is spare, geometrical, and of contemporary design, is more suitable for display type, such as that used in headings and titles, than for body copy.

Characteristics of Type

Type is measured in *points* and *picas.* The point is used to measure the height of a letter; 1 point is 0.0138 in. or approximately 1/72 of an inch. The pica is used for horizontal measurement of type. There are 12 points to 1 pica and 6 picas to 1 inch. Typefaces come in sizes from 6 to 72 points, with a complete font in each size. A font is an assortment of type in one size and color. Variations of a font may be available in light, bold, extra bold, expanded, and condensed.

Fonts consist of letters that are capital, also called caps or uppercase, and small, or lowercase. The term *case* originated with early compositors, who kept their type in different type cases. Fonts also include small caps, which are smaller letters in the shape of capitals.

In lowercase letters, the upper stroke (as on the letter d) is called an *ascender,* and the downward stroke (as on the letter p) is called a *descender.* Since uppercase letters are always on the line, they do not have ascenders or descenders.

Some typefaces have a crossline, or horizontal line, at the end of the main stroke of their letters. These crosslines are called *serifs* and the typefaces are serif fonts. Typefaces without serifs are referred to as *sans serif* fonts. The main portion of type is referred to as the *body.*

Typesetting, composition, and make-up are terms that refer to the process of setting and arranging type. The earliest typesetting was done by hand. Machine metal typesetting was developed in the late nineteenth century, when linotype and monotype were invented. Intertype and Ludlow were introduced in the early 1900s, and photocomposition became available in the 1950s. Special computers for typesetting were developed in the 1960s and 1970s.

Type Specification and Copy Fitting

Proper type specifications should include the following detailed information: the name of the type; its weight (light, condensed, extended, etc.); whether it is roman or italic; if it is all caps, small caps, or upper and lowercase; the size of the type (in points); the amount of space (called *leading*) between lines; the width (in picas) of the desired line; and whether the copy should be *justified* (i.e., flush right) or ragged right.

Copy fitting, or type calculation (often called *specking,* an abbreviated form of specifying), is a study in itself. It means to specify type accurately. The designer must refer to a comprehensive book of type specimens and, often, to books on custom-styled hand or photo lettering. These books are available from typographers.

Accurate composition depends on correct copy preparation and accurate markup instructions to the typesetter. Here are some of the rules:

1. All copy should be typed double-spaced on one side of an 8 ½″ x 11″ page. Copy should be marked up with colored pens or pencils to enable the typesetter to see instructions easily. All instructions should be specific. If necessary, use standard proofreaders' marks (found in most dictionaries under "proofreaders' marks").

2. The layout should accompany the copy so that the typesetter can get a clear picture of the job.

3. It is essential to check and edit copy *before* it is set. The typesetter sets the copy exactly as it is written. Spelling, punctuation, capitalization, and grammar should be carefully checked beforehand.

Press Type and I.N.T.® (Image N' Transfer)

Ready-to-use press type is available under various brand names. Type, symbols, ornaments, and simulated body copy are

---

® Industrial Graphics Division 3M.

printed on transparent sheets that can be transferred to most smooth surfaces by burnishing. Press type is usually produced in black and white. For permanent signage, adhesive-backed plastic lettering is available in many sizes.

There is a graphics system called I.N.T., which is a method whereby designers can create original art and type that can be transferred to packages. It can be prepared in a wide range of colors. This method is widely used today for the preparation of package comprehensives (commonly called *comps*). For more detailed information about this system, contact your local art supply store or manufacturer.

COLOR

Color as an aspect of design is becoming an increasingly visible part of our lives. Whether the medium is packages, printed matter, photography, movies, or television, color images have a greater impact than black-and-white. Color represents objects, scenes, and people with almost complete fidelity. In design, color can suggest abstract qualities, such as moods, temperament, warmth, coolness, and danger.

Color gets our attention. And when we first look at something, it's color that can create a pleasant, a shocking, or some other first impression. Color also has the psychological advantage of fixing visual impressions in memory and stimulating interest. Finally, color can add prestige to a package or advertisement. All three general reactions to color—attention, interest, and prestige—constitute part of its sales value.

We are taught from childhood to make certain associations with certain colors. For example, reds and oranges symbolize warmth, passion, war, danger, and a host of ideas connected with action and life. Blood and fire are red; our source of life, the sun, appears as a circle of red, orange, or yellow. Blues, on the other hand, symbolize ice, snow, and water. Almost all the colors of winter are tinged with blue. White is so closely associated with snow that it usually suggests coolness.

The basic suggestion of warmth or coolness is widely used in design. A picture of an ice bucket looks cooler rendered in cool whites or greenish blues than in oranges or reds. An electric heater would be rendered in reds, oranges, or yellows to suggest warmth and comfort.

Other abstract impressions can be suggested by color. Purity can be conveyed by white, light blues, pale greens, and other tints associated in our minds with things that are pure, such as water, snow, and blue sky. Sky blue frequently suggests serenity and peace. Hot pinks, reds, and yellows may suggest joy, gaiety, or the festive character of parties, celebrations, and parades. Mystery seems to lurk in soft, dusky hues. Deep reds, purples, and gold suggest riches and quality.

Of course, people's experiences differ, and we cannot be sure that a given color will suggest the same quality to all beholders. We know, however, that the distinction between warm and cool colors is relatively constant.

Designers can use the fact that people react to color. In most people, reaction to color is a pleasurable experience; it seems that a love of color is an inborn human trait. Colors that are both brilliant and brightly illuminated give us so much to see that we actually become keyed up, excited by the challenge to our faculties. Our response to color is similar to our response to music. Brilliant reds and oranges suggest loud, lively music. Soft hues remind us of slow, peaceful passages.

The Science of Color

There are many theories about the nature of color; they may be divided into three general groups.

1. Theories developed by physicists deal with the phenomena associated with light.

2. Theories developed by painters are based on the characteristics of pigments.

3. Theories suggested by psychologists relate to color sensations transmitted from the eye to the brain.

For the designer, the pigmental theories are the most important, although it is useful to be aware of the others as well. It is worth noting that none of these theories has been proven scientifically.

The color that is used in printing is pigmental color, that is, a form of paint. The difference between paint and printer's ink lies in the vehicle with which the coloring matter is mixed. The principal vehicle for oil paint is linseed oil; the vehicle for printer's ink is varnish.

Pigmental color differs from color produced by light in a significant respect: It is based on a different set of primary colors, that is, colors from which other colors can be produced. To appreciate this fact, it is necessary to know something about the science of color.

When a beam of sunlight is broken up by a prism, forming a beam of color called the *spectrum,* we see six major colors blending into one another: violet, green, blue, yellow, orange, and red. This classification is arbitrary. Sometimes a seventh color, indigo, is identified, located between violet and blue. A trained eye can recognize other intermediate colors, such as blue-green and yellow-green, at the points where one major color blends into its neighbor.

The colors in the spectrum are based on three primary wavelengths: red, green, and blue-violet. All the other colors are produced from combinations of these. In contrast, in pigmental color the three primary *hues* that are combined to produce all the others are crimson (magenta), yellow, and blue (cyan). When the three pigmented primary hues are mixed, the result is an approximate black.

Hue may be described as the quality of a color. When we distinguish among red, blue, and yellow as colors without any qualifying terms such as light, dark, weak, or strong we are classifying colors by their *primary hues.* These hues cannot be produced by mixing other hues. The *secondary hues* are orange, green, and violet. They can be produced by mixing equal parts of two primary hues. The *intermediate* hues are greenish-yellow and yellowish-green, bluish-green and greenish-blue, violet-blue, and bluish-violet, violet-red and reddish-violet (purple), orange-red and reddish-orange, yellowish-orange, and orange-yellow. These hues are produced by mixing unequal parts of two primary hues or of one primary and one secondary hue. As the proportions of the mixture are varied, the color inclines toward the hue that is used most. Additional hues—subdivisions of the intermediate hues—can be created, up to the point at which even a trained eye can no longer differentiate among them.

Value in a color is its degree of lightness or darkness. Black has the lowest possible value; white has the highest. In considering colors in terms of their value, we find that different hues differ in value. For example, if we place a pure yellow beside a pure blue, the yellow appears lighter than the blue. The yellow is more luminous (in fact, yellow is the most luminous color in the spectrum). The value of a color can be varied by adding white or black in different proportions. A hue that has been lightened by the addition of white is a *tint;* when it is darkened by the addition of black, it becomes a *shade.*

Chroma, or intensity, is the quality by which we distinguish strong colors from weak ones. Chroma can be defined as the purity of a color, that is, the extent to which it is free from neutral grays. A pure color cannot be toned down using gray or black alone. To weaken a pure color it is necessary to mix it with its complementary color. For example, when a small amount of green is mixed with red, the effect is usually pleasant, and the toned-down red has more color quality than would be obtained if the pure red were mixed with gray or black. As a rule, when a hue is mixed with a tint or shade of its complementary hue, the resulting tones are subdued, pleasant, and close to the original color.

Color harmony results from specific relationships among colors. There are three methods for achieving color harmony:

1. Use of various tones of the same hue, all in the same segment of the color wheel (*monochromatic harmony*)

2. Use of hues that are closely related but not identical

3. Use of complementary colors

Attempts have been made to determine arithmetically the proportions of various colors that must be used to achieve harmony. This is known as *color balance.* The basic principle of color balance is that the brighter and stronger a color is, the smaller its area should be. This should be true for all color combinations. In recent years, however, the use of newly developed luminous and fluorescent colors has resulted in combinations that may satisfy mathematical formulas yet do not achieve harmony. Both scientists and artists who study and work with color would agree that no formula can replace good taste and judgment. (Note, that is what designers are trained for.)

Color Matching Systems

To prepare package comps, designers can use any of the several color matching systems that exist. Color matching systems are commercial products available in art supply stores. They may take the form of adhesive-backed films, papers on coated and uncoated stocks, or screened colors. These materials are well suited to the preparation of attractive comps.

PHOTOGRAPHY

Photography probably began in the mid-1700s with the discovery that exposure to sunlight causes silver nitrate to turn dark. As early as 1700, scientists had discovered that when light rays converged on a small hole in a darkened box (known as a *camera obscura*) they carried to the inside wall of the box a soft image or reproduction of the view as seen through the hole. When simple lenses were substituted for the hole, the picture became clear and sharp.

In the early 1800s the sciences of chemistry and optics joined forces. In 1835 Louis Jacques Mandé Daguerre discovered a means of developing photographic plates coated with silver iodide by means of mercury vapor. By 1839 he had perfected a way in which to "fix" these images permanently on silver plates with a solution of sodium thiosulfate. Exposures required several minutes, but the quality of the pictures produced in this way, known as *daguerreotypes,* was exquisite (Fig. 2-11).

Optics and photographic plates were much improved by the late nineteenth century, and new, more reliable photographic papers were manufactured. But popular photography really arrived in 1884 with George Eastman's invention of flexible roll film to replace glass plates. In 1888 Eastman introduced a simple box camera (Fig. 2-12), making photography accessible to anyone who was attracted to this new and intriguing hobby.

Photography has come a long way since the days of the box camera. In this century color photography has completely changed the way we see the world. It has brought about changes in design, fashion, advertising, entertainment, education, and recording of current events.

In the United States, approximately 60 million cameras are in use, and more than 2 billion color and black-and-white photographs are made yearly. The variety of cameras and films on the market is

Figure 2-11. Daguerreotype from 1865.

Figure 2-12. Photography of the past. *Left to right:* daguerreotypes, tintypes, early box camera.

enormous. Tremendous quantities of silver are used in manufacturing the film; thousands of accessories must be produced to meet the demands of amateur and professional photographers.

As a profession, photography is an exciting occupation. The professional photographer is a trained, creative artist, designer, illustrator, and technician. He or she must be versatile and flexible. The photographer must be able to adapt to shooting in a studio or on distant locations, working with models, props, and many different kinds of equipment. Photography, in short, is a demanding but rewarding career.

A professional photographer needs a good-sized studio with a darkroom and dressing rooms. The studio should be equipped with lights, studio cameras, and basic props. Most photographers use a large, stable 8″ × 10″ studio camera to achieve high levels of accuracy, quality, and fidelity. For similar quality but less costly processing, a 4″ × 5″ camera is suitable. For action, speed, and economy, twin-lens reflex cameras (2 ¼″ × 2 ¼″) or 35 mm single-lens reflex cameras are popular. Polaroid or other instant print cameras may be used for test shots.

For the designer, photography is a form of illustration. It offers both fidelity to the subject and a creative, experimental quality. Photography can also be a time saver. It is suitable for product shots as well as for people, animals, and landscapes. It can be done indoors or outdoors under almost any light or weather conditions.

At times designers take their own photographs; at other times they work with professional photographers. It depends upon the skill and training of the designer in the art of photography. In any case, designers need to be acquainted with the work and equipment of the photographer. As such, photography is taught in all major art schools and is a required course for designers. Both the techniques of picture taking and darkroom technology must be learned.

A Typical Photography Project

Suppose you have to design a large package for a doll. In order for the product to sell, its package design must have absolute fidelity; that is, it must reproduce the product exactly. Rather than using an illustration, you therefore choose to have a professional photograph a little girl playing with the doll.

The first step in carrying out the project is to use markers to prepare a layout showing how you would like the final picture to look. Next, you need to have a preproduction meeting with the photographer. At this meeting you present your layout; describe your ideas about the set, lighting, mood, and the appearance of the child model; and indicate your budget. The photographer helps you clarify your ideas and solve any problems that might arise.

The next step is to interview (audition) models. Use models from a child model agency not amateur models, no matter how cute they are. Aside from possible legal problems, they lack the performance skills of trained models. Also, select the model's wardrobe and supervise make-up and hairstyling, paying special attention to nails and missing front teeth.

When the model and the photographer are ready, take a Polaroid test shot of your suggested layout. After studying the shot, see if you need to do several variations. Remember that film is cheap and time is money. Let the photographer shoot as many rolls of film as are needed to cover every possible situation and angle. Sometimes an unexpected angle will result in the best picture.

Since most photographers are working against time, you must make fast and correct decisions during a shooting session. The session should not take more than an hour or two. You will get the "chromes" the next day or earlier. *Chromes* are large transparencies (4″ × 5″, 8″ × 10″, or even larger).

Now comes the difficult task of selecting the right chrome. With 35 mm film, you get fifty or sixty possibilities; with 4″ × 5″ film, you usually get only 10 to 12 chromes (Fig. 2-13).

Once you have made your selection, examine it thoroughly over a light table, using a magnifying glass. This is the time to make changes on the photograph. If you notice, for example, that the model's hair appears too light, it can be corrected by an expert retoucher. Although there are some limitations, a skillful retoucher can solve a lot of difficult problems. The colors on chromes or transparencies can be corrected, altered, or changed from light to dark; colors can be added or removed; and patterns can be changed.

First, however, if the transparency is a 35 mm slide or a 2 ¼″ × 2 ¼″ chrome, a duplicate (*dupe*) must be made in a larger size, usually 4″ × 5″ or 8″ × 10″, to facilitate retouching. The retouching process basically consists of bleaching areas to be lightened and adding dyes to create deeper tones. Once the original emulsion has been removed from the chrome, it is impossible to replace it. Most retouchers, therefore, prefer to work on dupes.

Figure 2-13. The designer will choose the appropriate chrome for the package. *(Photos courtesy of Ideal Toy Corporation.)*

It is possible to make *assemblies,* that is, combinations of several silhouetted chromes that are pieced together to form a single large chrome. The preparation of these assemblies requires considerable technical skill and is quite costly, but the results are usually excellent.

Special Effects: Dye Transfers and C Prints

Certain design problems cannot be solved or corrected by a retoucher fixing a photograph. Retouchers cannot create effects such as a person flying through the air or a downtown view of New York City in the middle of a desert. For such effects a special print called a *dye transfer* is needed.

Before discussing dye transfers, however, it is important to be aware of an important distinction made in art and design between *reflective art* (paintings, illustrations, photographic prints) and *transparency art* (slides and chromes made on colo-positive films such as Kodachrome or Ektachrome). Reflective art more closely approximates the appearance of the printed page and therefore affords more accurate reproduction. In transparencies, several hundred times more light passes through the highlights than through the shadows. As a result, transparencies have considerably more detail and brilliance than reflective art and are often preferred for fine color printing.

Dye transfers are the finest-quality prints available for reproducing an image, either from a transparency or from reflective art. The highly controlled technique and dyes used in the process permit the closest possible match to the original art. It takes several days to produce these prints, which are costly but remarkably accurate in both color and detail. Using dye transfers, a skilled retoucher can achieve unbelievable effects.

More modest in price are Ektacolor or *C prints,* which can be produced in a few hours. The quality of C prints varies. Nevertheless, they are widely used for packages, layout, displays, and other visual aids. Still more economical are color photostats, which are extremely useful for simple basic presentations. A good Xerox-type duplicator with laser (color prints) can be used to create many striking graphic effects inexpensively.

There are several systems for producing full-color comprehensives for packaging. For information about these systems, check the trade ads in graphic design publications and visit trade shows, where you can see them demonstrated by the manufacturer or the service organization.

Photography versus Illustration

The relative merits of photography and illustration are often debated by designers and their clients. It is true that photography suggests fidelity and truth, whereas illustration tends to be more individualistic. It is, in effect, a direct statement by the illustrator. The fact is, each job presents its own specific problems. The designer should solve these problems by using the medium that is most appropriate in the particular situation, keeping in mind the standards of good taste and design and the requirements of the client.

Photography as a Presentation Tool

Suppose you have been asked to design an elegant package for a light summer cologne. You think of a lovely view; perhaps a close-up of grass, plants, and flowers that suggests a particular mood. How can you convey your idea to the client?

First, take your camera and try to find a scene like the one you have imagined. Take a photograph of it using good-quality Ektachrome film, which can be processed quickly. Next, project your slides directly onto a blank carton, using the projector to reduce or enlarge the size of the image. This is an excellent, time-saving way to come up with visuals for a preliminary design presentation. Your client will be impressed by the promptness and clarity of your presentation.

PHASES OF A DESIGN PROJECT

Most design projects are presented and costed out in four phases. The first phase involves *gathering information* (i.e., research). At this time, ask the following questions: Where will the product be sold? Who will buy it? How much will it cost? These are basically marketing and production issues of concern to the designer. In-store research can be useful, along with information gleaned from books, periodicals, and swipes.

The second phase is the *development of preliminary concepts.* As shown in Figure 2-14 (the development of a logo, i.e., corporate identity), small *thumbnail sketches* or "roughs" can be prepared using a 2B pencil, a pentel, or markers. Make at least 12 to 15 sketches or roughs. Thumbnails are usually in black and white; roughs are in color on a 14″ × 17″ layout pad, a useful size. Clearly indicate type styles and logos. Make sure your roughs and sketches are neat and clear, and keep them in your portfolio for future reference and for use as presentation material.

The third phase is *preparation of the comprehensive* (comp). Figure 2-15 shows a typical comp showing very sharp work. This phase includes determining the structural design of the carton, box, model, or prototype. Be sure to consider potential production and manufacturing problems. Many great designs have gone down the drain because the production and manufacturing costs would have been prohibitive.

Before you construct a final comp, prepare a *mock-up,* a rough structural device that serves to test the viability of the design concept. For a small carton, three-ply bristol board or 24 pt board is most suitable. Boards sometimes warp. You can avoid this problem by attaching (*mounting*) art to the carton. For mounting artwork, use rubber cement or spray. For glueflap or cutout type, use white glue (e.g., Sobo or Elmer's).

If you are planning to do artwork directly on the board with water-based colors, be sure to try out colors, washes,

Figure 2-14. Preliminary concepts. Sketches for a corporate identity logo by Lisa McGowan.

Figure 2-15. Comprehensives (comps) are full-scale, flowless structural and graphic presentations. *(Photo courtesy of Ideal Toy Corporation.)*

and other techniques on a separate board. Hand-lettering should be complete. Transfer type (I.N.T.), embossing, and diecuts can be used for this purpose.

The final phase of the project is *preparation of the production mechanical,* which incorporates all final art, photography, illustration, and type. If the packaged product is to be sold in supermarkets, you will need to have a Universal Product Code (UPC) symbol printed on the package. An optical scanner can "read" the UPC symbol into a computer, which will indicate the price and provide a printed tape itemizing the products purchased and totaling the bill. The UPC symbol is about 1½ sq. in. and consists of 30 vertical dark lines with 29 spaces and a 10-digit series of numbers. The first five numbers designate the manufacturer; the second five identify the product and the package size.

COMPUTERS AND THE PACKAGE DESIGNER

Since the introduction of the computer, typesetting techniques, composition, and make-up have been revolutionized. Several computer-based typesetting systems are available, including the Videocomp, Harris, Mergenthalen, and Linotron series. Compugraphic Video Setter and Autologic APS are capable of very high speeds and in some cases can create complete pages of text with headline material in position.

Word processors are essentially typewriters connected to some form of recorded medium. Today most word processors are video-based and can carry out composition and make-up functions similar to those offered by typesetters. The word processor's recorded medium permits typists to "capture" keystrokes for editing and correction.

Newly introduced copier/duplicators use digitized methods to produce pages for high-speed reproduction. In recent years digitized methods (i.e., a method producing high-speed repros) have been used for platemaking, scanning a page pasteup and then using lasers to produce the plate.

Computer-Aided Package Design

Computer-aided package design systems are the most significant development in contemporary packaging design. They enable the designer to create or alter package concepts in minutes. A computer image-producing system can replicate any two-dimensional object on the screen, allowing the designer to experiment with various portions of the image. The designer can alter colors, intensify background tones, add or remove shadows. Text or images can be squeezed, stretched, rotated, superimposed on another image, zoomed in, or moved to other parts of the screen. New components can be added. Typefaces can be adjusted, proportioned, moved, or erased. The options are limited only by the user's imagination. Once a designer has used the system, the most common reaction is "How did I ever get along without it?"

The following is a description of how computer-aided package design works. Using an attached flatbed scanner or video camera, the computer can "grab" images from drawings, photographs, printed materials, transparencies, and even three-dimensional objects. When an image is fed to the screen, the designer manipulates it. In Figure 2-16, the designer is developing a logo. Since the computer design system has 1800 typefaces, many different possibilities for the logo can be created in a matter of minutes. Then, also in a short time, several logos can be chosen and further developed (Fig. 2-17)—a procedure that used to take days or even weeks.

When the concept is completed (Fig. 2-18), the designer may choose among several options to reproduce it: Polaroid print, a 35 mm transparency, or an ink-jet or thermal-printer copy.

The computer can also simulate the environments in which designs will appear. Suppose you want to know how a design will look in a retail environment. You can input the interior of a supermarket, using a slide, and superimpose your design. The result is a *planograph* "store shelf" view, in which several packages appear in a row as they would on a store shelf. You can also create a supermarket display in which the impact of the package can be

Figure 2-16 (opposite page). Designing with the computer by Karen Whitlock. Phase 1 components.

kidamins
kidamins
kidamins
kidamins
kidamins
kid'amins
kid'amins
kid'amins
kid'amins

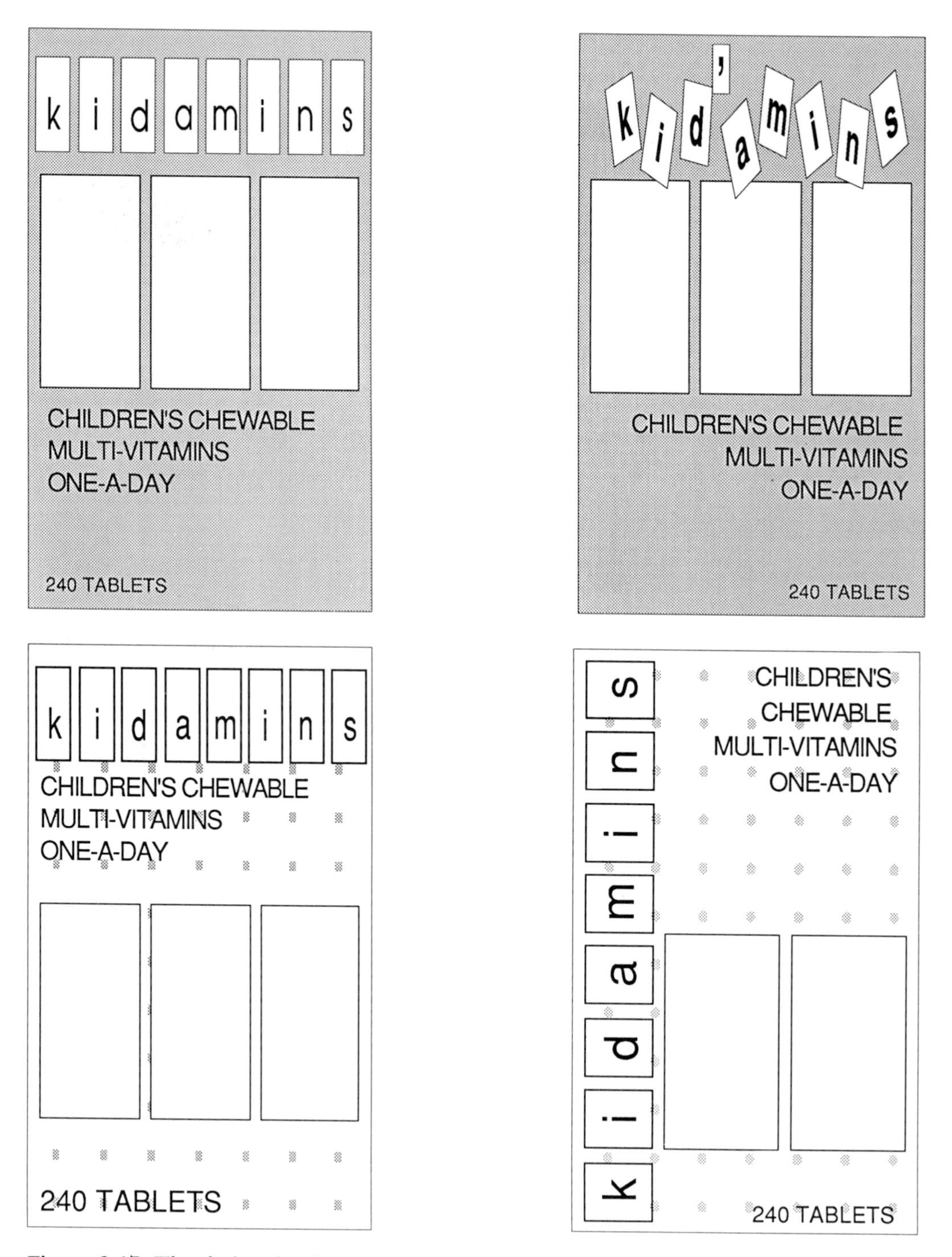

Figure 2-17. The design development of the package with the computer by Karen Whitlock.

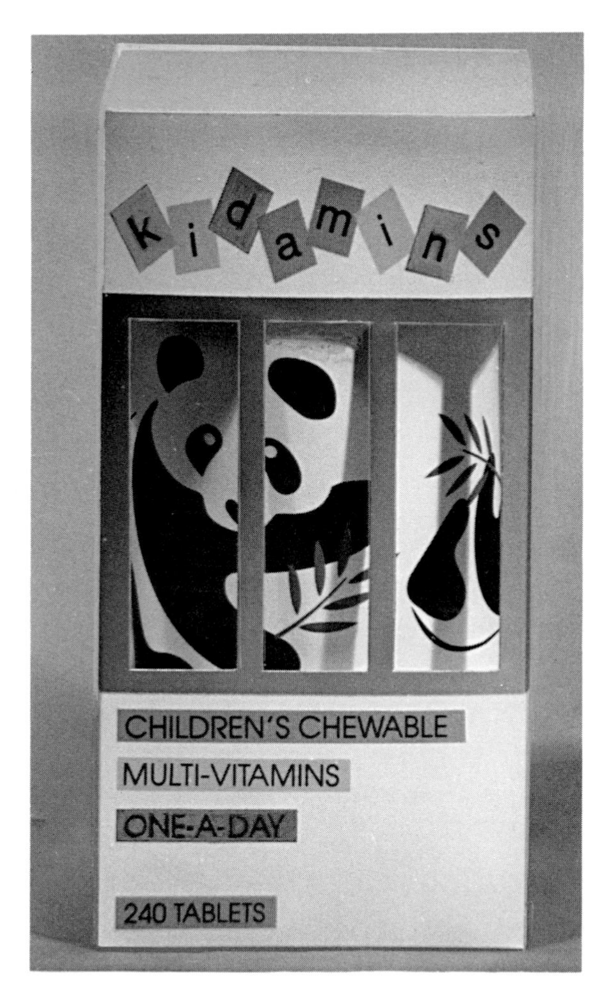

Figure 2-18. The final comp by Karen Whitlock. *(Photo by Miles David Sebold.)*

compared to that of competitors' packages.

Some of the most popular computers used in packaging design are Apple Macintosh, Lightspeed, IBM PC and compatibles, and Aesthedes. Computer-aided design (CAD) programs are available in a wide variety of price ranges and levels of sophistication. Whatever input device you use, computers greatly reduce the amount of time required for revisions and modifications of drawing board art.

PREPARING YOUR PORTFOLIO

Your portfolio is your introduction to the job market. It should therefore display your skills, your creativity, and to a certain extent your personality. Putting together a good portfolio, one that will sell your talents to prospective employers, is both an art and a science.

As a package designer, your portfolio is somewhat different from other art portfolios because most of the work is three-dimensional. You should take slides (35 mm) of your work and choose 12 to 18 of the best ones for your presentation. For especially outstanding work, present six or eight transparencies of different views. These should be mounted on black mats, which are available in photographic supply stores.

In addition to slides of finished work, include at least six excellent marker renderings of diverse subjects—people, objects (cameras, toys, etc.), and of course packages. Also include hand letterings, logo designs, one good mechanical, and some of your thumbnails and roughs. If you have other skills, such as retouching, airbrush art, or photography, show samples. Do not include too many items of the same type. When you present your work, bring along a resume and a well-designed business card. Almost everyone has a business card. But as a designer, yours is especially important. Design it yourself, avoiding complex designs, textures, and colors.

Carry a practical portfolio with individual art matted separately. Do not use a zippered portfolio with acetate pages. The rigid, case-like portfolio is most practical for the packaging designer.

The average interview for a newcomer lasts only 15 or 20 minutes. Therefore, it is important to be brief and articulate in talking about your work.

PROJECTS

1. Get a sketchbook and number all the pages. Draw anything and everything you see around you, using all available media. Take your sketchbook with you wherever you go. Do not remove any pages.

2. Develop your own corporate symbol. Your logo must be suitable for stationery, business cards, products, packaging, and signage. It should be in color, but it must also look good in black and white. To get some ideas, study and analyze some of the better-known corporate logos.

3. Design a series of packages: a cereal box, a simple beach toy, a shampoo for blonds, and a country-and-western album cover, the basic color scheme of which is yellow, ranging from cadmium to lemon yellow. You could use yellow for an overall effect (monochromatic rendering with contrasting logo and lettering), or you could print a black-and-white illustration or photo on yellow stock. A bright yellow hand-lettered logo could serve as the main design element, or you could emphasize "symbols of yellow" such as the sun, the sky, or plants.

4. Use your camera in designing a package for a preschool toy. Use two small children as models. Be sure that they play with the toy and look into the camera.

SUGGESTED READINGS

Briggs, John R. *Basic Typography*. New York: Watson-Guptill, 1972.

Bruno, Michael H. *Pocket Pal*. New York: International Paper Co., current edition.

Craig, James. *Designing with Type*. New York: Watson-Guptill, 1980.

Donahue, Bud. *The Language of Layout*. Englewood Cliffs, NJ: Prentice Hall, 1978.

Edwards, Betty. *Drawing on the Right Side of the Brain*. Los Angeles: J.P. Tarcher, 1979.

Gates, David. *Graphic Design Studio Procedure*. Monsey, NY: Lloyd-Simone, 1982.

Gates, David. *Type*. New York: Watson-Guptill, 1973.

Itten, Johannes. *The Art of Color*. New York: Van Nostrand Reinhold, 1962.

Kerlow, Isaac V. and Judson Rosebush. *Computer Graphics for Designers and Artists*.

New York: Van Nostrand Reinhold, 1987.

Leach, Mortimer. *Lettering for Advertising*. New York: Van Nostrand Reinhold, 1968.

Nesbit, Alexander. *The History and Technique of Lettering*. New York: Dover, 1957.

Nicolaides, Kimon. *The Natural Way to Draw*. Boston: Houghton Mifflin, 1975.

Reedy, William A. *Impact: Photography for Advertising*. Rochester, NY: Eastman Kodak, 1973.

Schiffman, Leon G., et al. *Consumer Behavior*. Englewood Cliffs, NJ: Prentice Hall, 1978.

Troise, Emil and Otis Port. *Painting with Markers*. New York: Watson-Guptill, 1972.

3

PAPER, BOARD, AND STRUCTURAL DESIGN

Types of Paper and Board / Working with Paper and Board / Folding Cartons / Set-up Paper Boxes / Corrugated Containers / Point-of-Purchase Displays

PAPER gets its name from papyrus, a reed whose fibers were used by Egyptians to make paper beginning around 2500 B.C. (Fig. 3-1). About A.D. 105 the Chinese invented paper as we know it today, and their art of papermaking spread to other parts of the world. In A.D. 795 a paper industry was established in Baghdad. In the wake of the Crusades and the Moorish conquest of Spain and North Africa, papermaking spread throughout the civilized world.

Figure 3-1. Egyptian scribe using paper made from papyrus.

The first paper was made by pounding the inner bark of the mulberry tree into sheets. Later it was discovered that better paper could be made by pounding rags and hemp into pulp. For centuries paper was made by hand from rag pulp. Although that was a slow process, it was enough to satisfy the demand for paper—until the invention of printing.

With the invention of printing from movable type by Johann Gutenberg of Mainz, Germany, in 1450, the demand for paper increased. To satisfy that demand, in 1798 Nicolas Louis Robert invented a machine that made paper in continuous rolls rather than sheets. By 1803 the Fourdrinier family had taken over the financing and development of the machine. The machine is still referred to as a Fourdrinier.

The process for making paper continued to improve. In 1840, Friedrich G. Keller in Germany invented a way to grind logs into a fibrous pulp known as groundwood pulp. In 1867, Benjamin C. Tilghman in America discovered how to separate the wood fibers by dissolving the wood in a solution of sulfurous acid. By 1882 the process of producing wood pulp had become similar to that used by modern paper mills.

Today, there are three major processes for making pulp from wood. They are the *sulfate, sulfite,* and *caustic soda* processes. These are all cooking processes in which the pulp is washed, bleached, and cleansed of all foreign materials and then fed into a machine that will form the mass of fibers into a sheet. At that point the pulp can be colored with dyes.

Different types of paper can be made. When a smooth surface is needed for printing or writing, glue or starch is added. When a slick surface is desired, the paper sheet is covered with special clay or fine plastic coating material. Paper can also be coated by being impregnated with waxes and plastics. Finally, to increase the smoothness and gloss of its surface, paper is passed through a set of iron rolls at the end of the paper machine. This process is called *calendering.*

TYPES OF PAPER AND BOARD

Board is made by laminating layers of paper together. If the board is more than 0.0012″ thick (12 points), it is paperboard or cardboard. If it is between 18 and 24 points thick, it's the boxboard used in folding cartons. Illustration board is about 60 points thick. For displays or signs 80–100 point board is generally used. The Bristol board used for drawing, pen-and-ink renderings, and watercolor comes in 12, 15, 18, or 21 points; that is, 2, 3, 4, or 5 plies.

The paperboard used in cartons is specified according to the size of the carton or, more often, the weight of the item that goes into it. A glass bottle for 3.5 fluid ounces of a fragrance, for example, would take a folding carton of approximately 18 to 24 points. The paperboard's thickness is expressed in what is called caliper points. *Caliper* is given in units of a thousandth of an inch (usually written in decimals). (Since most papers are laminated or coated with other material, caliper points are rarely used today to specify weight.)

Papers Used in Packaging

Different types and grades of papers are available for different uses. The following are some basic papers used in packaging:

Unbleached kraft, or coarse brown paper, is the most economical and strongest packaging paper. It is used for wrapping paper, paper bags, and general packing purposes. Unbleached kraft can be laminated, coated, and impregnated with various protective materials such as plastics and waxes.

Glassine and greaseproof papers can be plain, printed, lacquered, waxed, corrugated, or laminated onto other packaging materials. The outstanding qualities of these papers are their ability to act as a barrier against water, vapor, and odors, and, especially, their resistance to grease. Glassine and greaseproof papers include pouches, bags, lined cartons, and envelopes. About 85 percent of them are used in food packaging.

Parchment papers are made by dipping sheets into a concentrated solution of sulfuric acid. The result is a tough, dense, translucent film that is sterile, free of fibers, strong when wet, and highly greaseproof. Parchment is an excellent liner or wrap for oily or wet items like butter, fish, and vegetables.

Tissue is used primarily as an inner wrap. It may have a hard or a soft surface, can be waxed or impregnated with plastic resins for strength, and is available in both translucent and opaque colors. Tissue is widely used by florists and in the hosiery and food industries.

Sulfites, clay, and chromecoats are used in printing, labeling, and decorative packaging. They are of two types, one with a flat or dull finish (coated and uncoated) and one with a glossy finish. Smooth high-gloss paper comes in brilliant white and colors; especially beautiful are the flint papers used for covering boxes and gift wrapping. The glossy-finish supercalendered whites (sulfites, clay, and chromecoats) emboss well and are excellent for high-quality printing.

Foils are produced by laminating metal foil to paper or by gravure-printing paper with metallic powders mixed with lacquers. They are available in a wide range of colors and finishes and can be beautifully embossed. Foils are used for speciality box coverings and overwraps.

Specialty papers are textured with flock, glitter, foam, and other materials. Iridescent and pearlescent papers are used in box coverings and platforms (inserts) for gift and luxury items.

Boards Used in Packaging

Boards account for more than 60 percent of the materials used in packaging, including folding boxes and set-up boxes (see later discussion). With the exception of the low-cost, plain chipboard, most boxboards are lined or laminated to another paper liner or liners. The inner board is usually 100 percent recycled fiber derived from low-grade papers. The outer or top liner varies according to need and quality. The basic boards used in packaging follow.

Plain chipboard, the lowest-cost board, can be adapted for special linings. It is not suitable for printing. Colors range from gray to tan.

White wat-lined chipboard has a white liner that can be adapted for color printing. It is used for high grade set-up boxes with a white liner.

Bending chip, the lowest-cost boxboard, is used in inexpensive folding cartons. It is usually light gray or tan but can be printed on in any color.

White-lined 70 newsback is a smooth board that is much whiter than most of the inexpensive grades of board. The back is usually gray. It is excellent for folding cartons, posters, displays, and die-cut items.

Bleached manila-lined bending chip is the same as bending chip except that the top white liner is of better quality.

Clay-coated boxboard is a smooth, white board with an excellent printing surface. It is used for cartons and displays and whenever high-grade multicolor printing is needed.

Solid manila board is available with a white liner and a manila back. It is used for all types of cartons that require durability and strength.

Extra-strength plain kraft-type boards are available with or without the white liner. They are used for hardware, automobile parts, housewares, toys, and other items that require extra-strength packaging.

Uncoated solid bleached sulfate is strong, white board that is plastic coated or waxed and hence water resistant. It is excellent under freezer conditions and is widely used in frozen-food cartons.

Clay-coated solid bleached sulfate offers excellent scoring and folding characteristics, is highly uniform, and provides a good surface for printing. Its appearance (dull to high gloss) depends on the coating process used. It is most suitable for quality goods such as cosmetics, gifts, and pharmaceuticals.

Clay-coated natural kraft is a strong, moisture-resistant board with a white printing surface. It is used in heavy-duty packaging, bottles, beverage carriers, and similar items.

Ovenable paperboard (for microwave ovens) is a paperboard with a heat-resistant coating that will withstand conventional oven temperatures and allow microwave energy to pass through it. Coated solid bleached sulfate is the most frequently used ovenable paperboard.

WORKING WITH PAPER AND BOARD

Although paper and board come in flat sheets or rolls, different things can be done with them to create packages and other three-dimensional objects.

Grain

In all papers and boards the fibers are aligned in one direction, called the *grain.* The paper or board will fold or score easily along the grain. If they are torn with the grain, their edges will be smooth; if they are torn against the grain, their edges will be ragged.

Scoring

Paper can be easily folded. To facilitate folding, a *score,* or crease, is made in the

Figure 3-2. Embossing plate. *(Photo courtesy of Helena Rubinstein, Inc.)*

paper or board. The tool used for scoring is a blunt-face (round edge) scoring rule (die). Scoring rules come in different widths for different thicknesses of paper.

When constructing a carton by hand, *never score the paper with a sharp blade.* A cut will crush the fibers and weaken the paper. Use a blunt edge, such as a paper clip, coin, or ballpoint pen, against a steel rule. Always bend against the score to produce an embossed, raised edge.

Embossing

Paper and board also lend themselves to *embossing*, the process by which a design or image is made to appear in relief on the paper or board (Fig. 3-2). Embossing can be superimposed on printing or done on blank paper for a sculptured, three-dimensional (*bas relief*) effect. The latter is known as *blind embossing*.

Embossing is achieved by pressing a sheet of paper between a brass female die and a male bed or counter, both of which are mounted in register on a press. Since embossing is a costly process, it is generally used for prestigious packages, cosmetics, gifts, stationery, and promotional materials.

Die-Cutting

Every paperboard product or paper, whether three-dimensional or flat, has a shape or form that is produced by die-cutting. The process of *die-cutting* involves creating shapes of many kinds, using cutting and stamping dies, from papers, board, and plastics.

There are three methods of die-cutting. *Hollow die-cutting* is done with a hollow die, which looks like a cookie cutter. This method is used almost exclusively for labels and envelopes.

Steel-rule die-cutting is used when close register is required. Steel rules are bent to the desired shape and inserted or wedged into a ¾″ piece of plywood. The multiple dies are locked up in a chase on a platen of the die-cutting press. Several

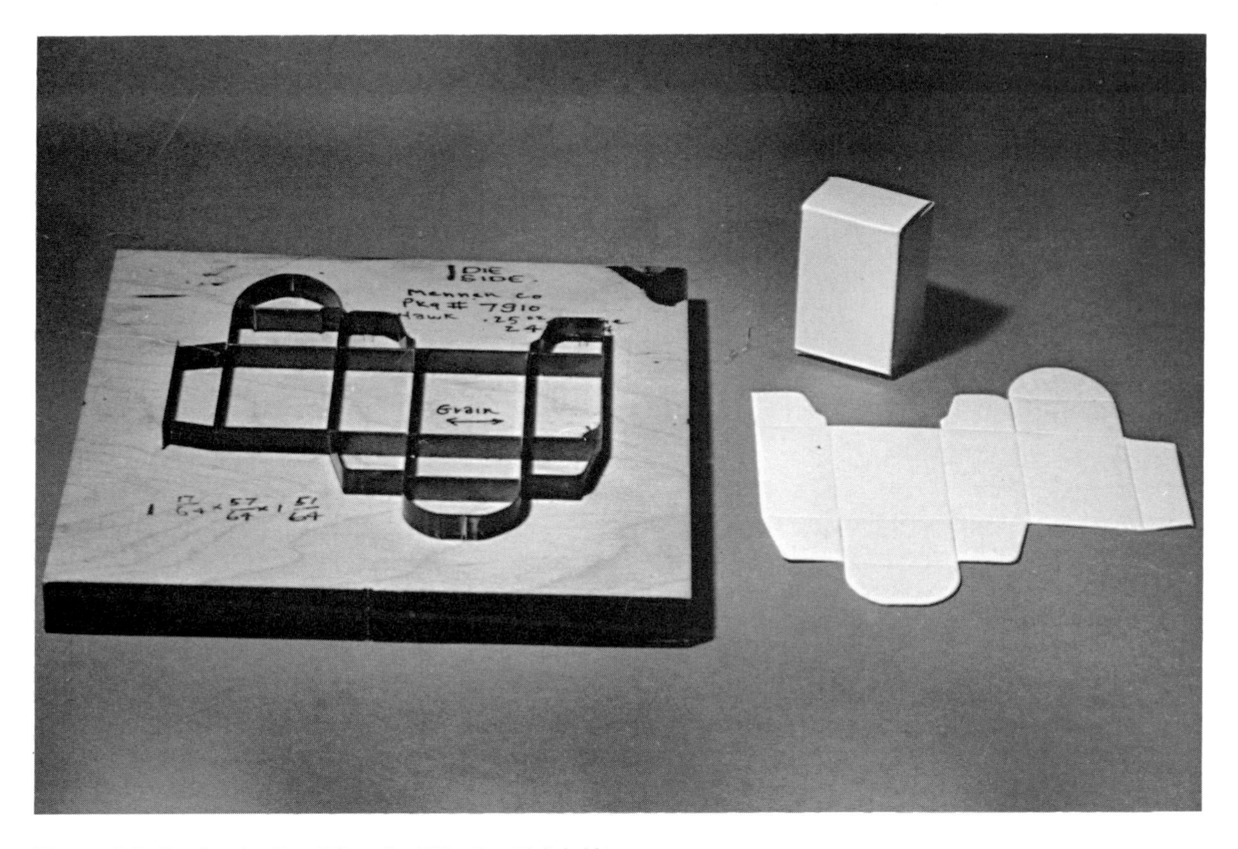

Figure 3-3. Steel-rule die. *(Photo by Miles David Sebold.)*

sheets can be cut at one time. A flatbed cylinder press can also be used for die-cutting. Figure 3-3 shows a steel rule die and the end product made from it.

The third die-cutting method uses *lasers*, which were invented by C.H. Townes and Arthur Schawlow in 1958. *Laser* is an abbreviation for light amplification by stimulated emission of radiation. The laser beam, which can be concentrated on a small point and used for manufacturing processes such as drilling, cutting, and welding, has changed many processes in manufacturing, communications, and medicine.

All types of materials, including paper, metal, plastics, and wood, can be die cut with lasers. Since the laser beam is extremely sharp and precise, the cutting is very accurate. Therefore, the resulting edges do not have to be finished in any other way, such as filing or buffing, as they do with other methods. Figure 3-4 shows a laser cutout.

Figure 3-4. Laser cut-out.

Adhesives

Paper and board packaging materials need to be joined or fastened together. Adhesives are used for those purposes. An entire industry is based on the production of glues, cements, gums, and hot-melt adhesive materials.

Designers need to be familiar with the adhesives used with different types of packaging materials, such as resin emulsion adhesives for coated boards and hot-melt adhesives for plastics and plastic films. They also have to be aware of government regulations, such as the federal regulations for the use of adhesives in food packaging.

Commercial and Industrial Packaging

Many types of adhesives are used in commercial and industrial packaging. Self-adhesive and pressure-sensitive labels, for example, use a synthetic latex that is suitable for both removable and permanent labels. A special resin-emulsion adhesive is needed for some clay-coated boards. Hot-melt adhesives are suitable for adhering plastic films. The adhesives used in industrial packaging may be reinforced, pressure-sensitive, and made from cloth, film, or foil.

Handmade Packaging

Handmade comprehensives should be joined together with white glue, which is a permanent adhesive. *Never use rubber cement for the glue flap* since it has a tendency to dry out. A fine coat of rubber cement, however, is suitable to mount fabrics on thin paper or board; spray glue will also work.

FOLDING CARTONS

In 1879 a Brooklyn printer, Robert Gair, was inspecting a printed seed package that had been inadvertently cut by an improperly positioned printing plate. It occurred to him that it might be possible to make a die press that could score and cut paperboard in one impression. Such a press would be suitable for cutting out carton blanks. Gair's idea was the birth of the *folding carton* (Figs. 3-5A and B).

Figure 3-5B. An early folding carton. *(Photo courtesy of Landor Associates.)*

Figure 3-5A. The first Ivory Flakes folding carton (1900). *(Photo courtesy of Landor Associates.)*

In 1987, folding paperboard cartons was a $3 billion industry. It then consisted of about 530 carton manufacturers with 752 plants that employed nearly 80,000 people. The plants used about 3.5 million tons of board to manufacture more than 250 million cartons a year.

Folding cartons are precision-made, low-cost packages supplied in *knock-down form,* also known as *blanks* (Fig. 3-6).

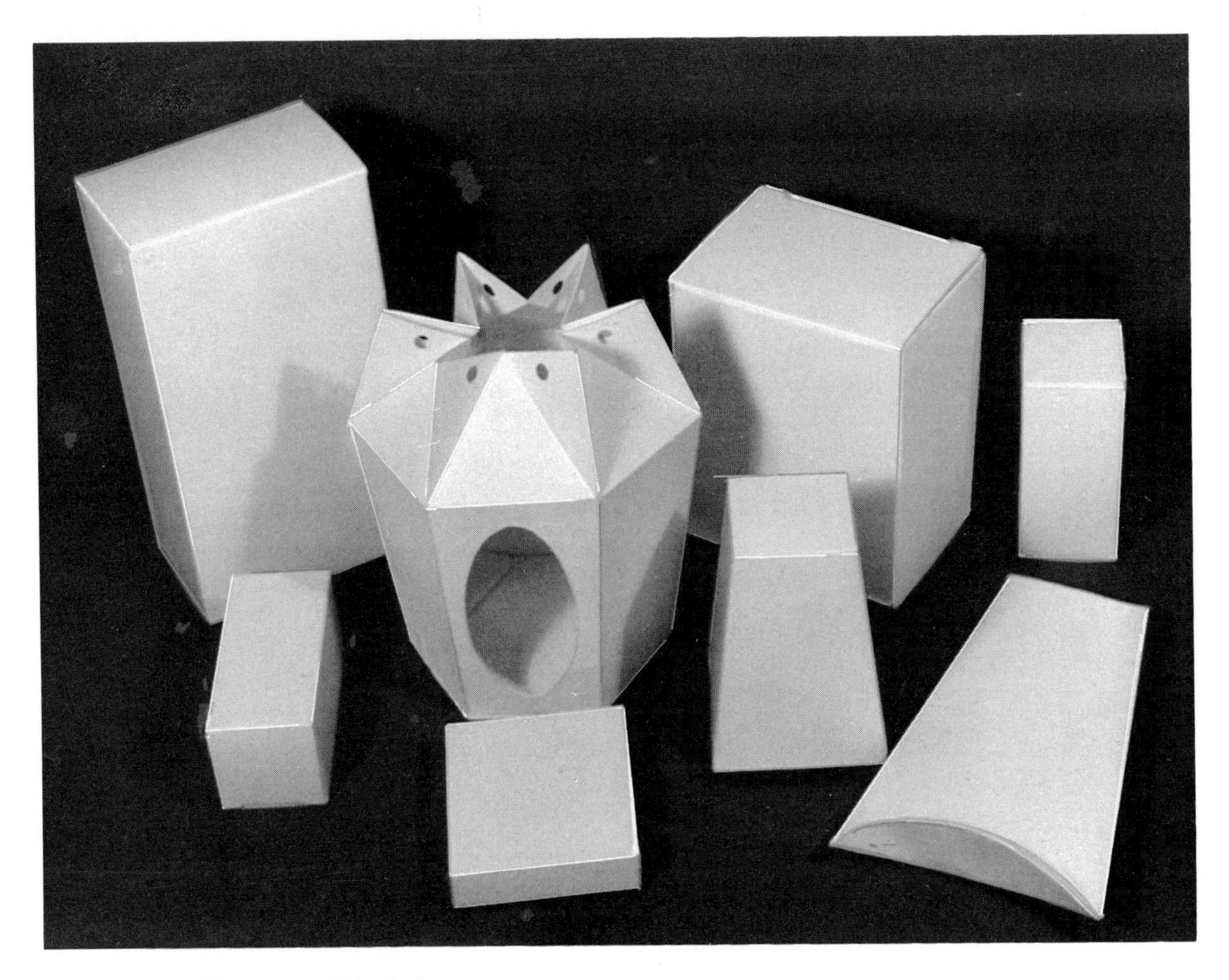

Figure 3-6. Folding carton "blanks."

When assembled, they become three-dimensional, rigid packages. They can be packed by high-speed automatic, semiautomatic, or hand-operated equipment. Knock-down containers lend themselves to various types of marketing and retailing systems, such as those for food, gifts, pharmaceuticals, cosmetics, toys, hardware, and housewares.

Types of Folding Cartons

Structural designers can choose from about 500 styles and variations of carton constructions, with more being added every year by skilled paper engineers. The styles and construction are determined by the product to be packaged and the type of filling operations that will be used. Filling operations are done on fully automatic filling equipment. There are several types of such equipment; the one used depends upon whether the item is bulky, liquid, powder, or granular.

Tray-Style Carton

One basic style of folding carton is the *tray* (Fig. 3-7). Figure 3-8 illustrates folding carton patterns. In one type of tray-style carton, solid bottoms are hinged to side and end walls. The sides and ends are connected by a flap, hook, locking tab, or lock. These cartons have a variety of cover and flap parts extending from the walls and sides of the tray. In another type of tray carton, two pieces, one slightly smaller than the other, form the base and cover of a two-piece telescoping box. Typical tray packages are cigarette cartons, bakery trays, ice cream cartons, pizza cartons, and garment carriers.

Tube-Style Carton

Another basic type of folding carton is the *tube* (Fig. 3-9). The body of a tube-style carton is a sheet of board that is folded over and glued against its edges to form a rectangular sleeve. It has openings on the top and bottom that are closed with flaps, reverse or straight tucks, and locks. Tube-style cartons give the product fully enclosed protection. They are therefore used to pack bottled

Figure 3-7. A tray-style carton.

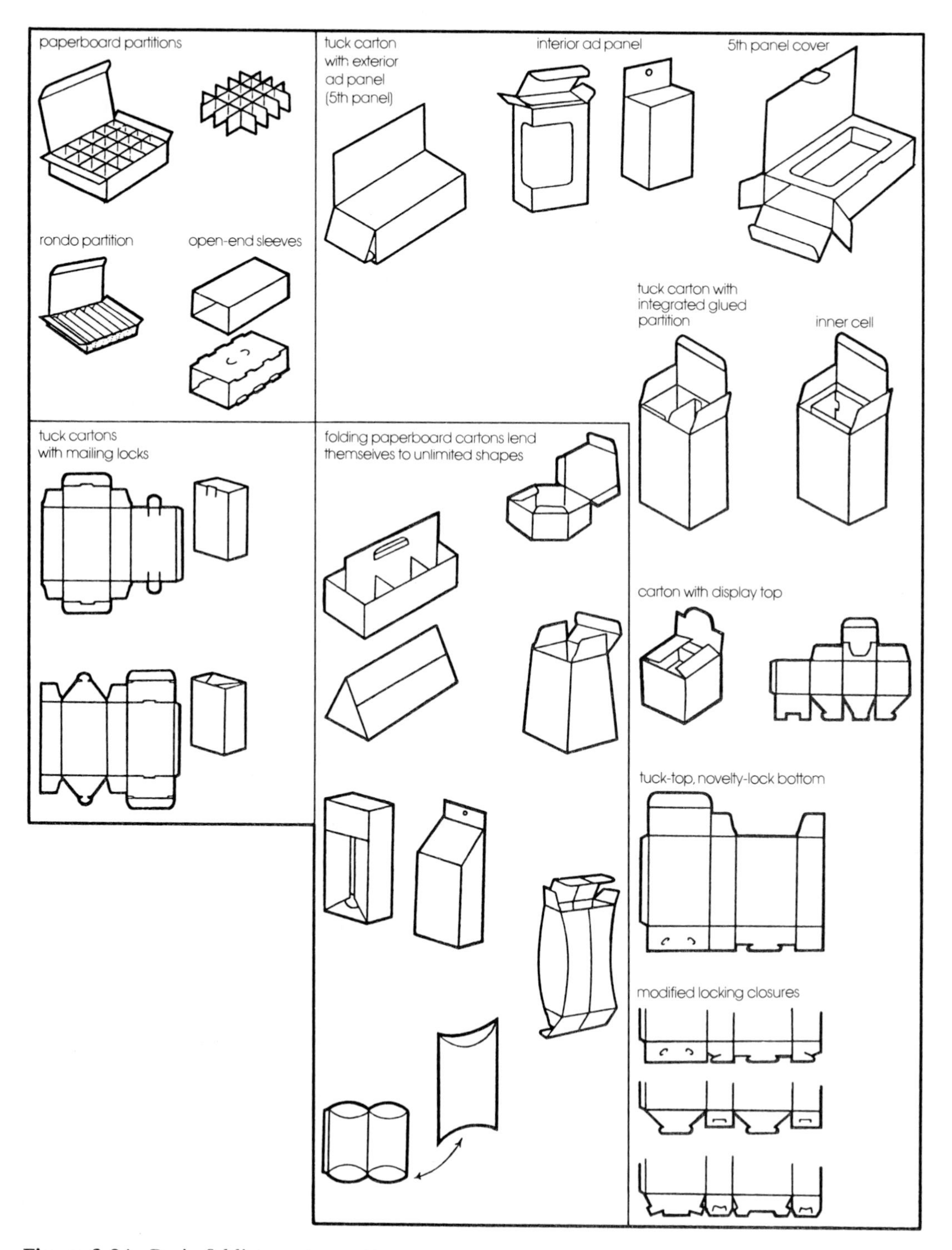

Figure 3-8A. Basic folding carton patterns.

Figure 3-8B. Basic folding carton patterns.

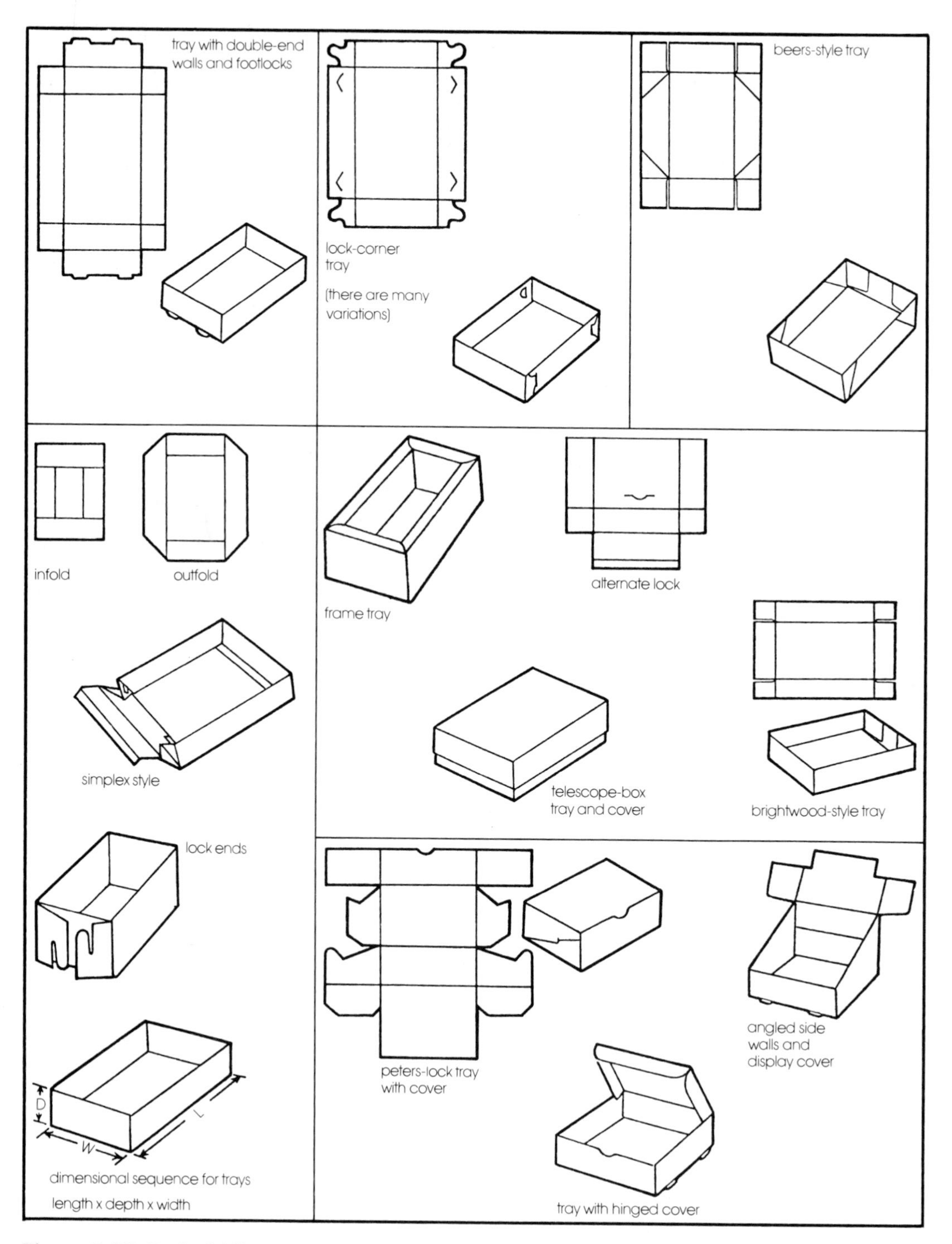

Figure 3-8C. Basic folding carton patterns.

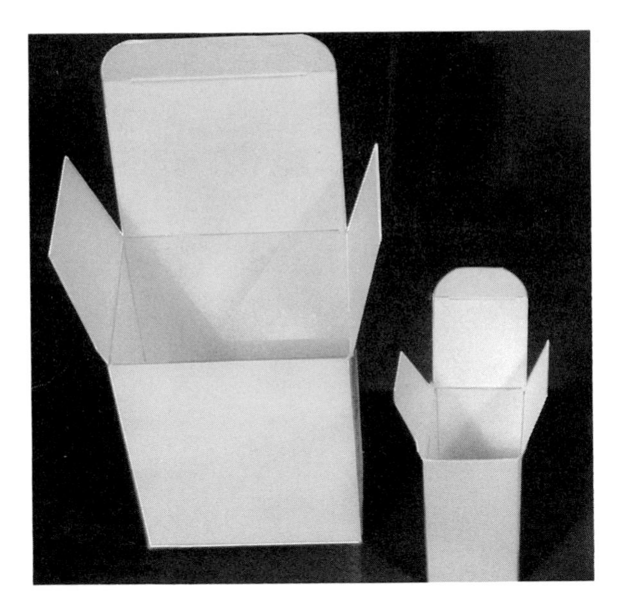

Figure 3-9. A tube-type carton.

products, cosmetics, and pharmaceuticals. Often windows are added, which enable purchasers to see the product. Many unusual tube styles are available, including contoured, triangular, octagonal, or even rounded. Figures 3-10 through 3-16 show different types of tube-style cartons.

Shrink-Wrapping

Another method of packaging toys, housewares, and contoured products is to shrink-wrap them. *Shrink-wrapping,* which involves sealing a layer of plastic around an object with the application of heat, is a good way to display as well as protect products.

Figure 3-10. Fast-food packaging. Lawrence McGarvey, designer. *(Photo by Miles David Sebold.)*

Figure 3-11. Cereal packages. *Left to right:* Marjorie Wood, Eliana Themistocleous, and Suzette Cascio, designers. *(Photo by Miles David Sebold.)*

Figure 3-12. Self-service folding cartons. *Left to right:* Lisa McGowan, Lawrence McGarvey, and Cindy Campo, designers. *(Photo by Miles David Sebold.)*

Figure 3-13. Pet food carton. Cindy Campo, designer. *(Photo by Miles David Sebold.)*

Figure 3-14. Die-cut, animated vitamin package for children. Susan Rajnert, designer. *(Photo by Miles David Sebold.)*

Figure 3-15. Die-cut "sleeves" for tissue carton. *Left to right:* Elizabeth Downey, Ginger Valentino, and Lorraine Casscles, designers. *(Photo by Miles David Sebold.)*

Figure 3-16. Premium packaging sleeves for milk or juice cartons. *Left to right:* Kim Findley and Donna Capetta, designers. *(Photo by Miles David Sebold.)*

Adaptations of Folding Cartons

Adaptations of the folding carton include the bag-in-the-box, boil-in-bag pouches, soups in pouches, and paper frozen-food cartons that can be heated in a conventional or microwave oven and used as a serving dish. Beverages such as milk and fruit juices are packaged in specially designed folding cartons that are lined with film and foil. This type of package is often referred to as *aseptic packaging.* (See also Chapter 4, The Age of Plastics, and Chapter 5, Flexible Packaging.)

Many innovative modifications of folding cartons are found in the institutional market for food products. One of these is a liquid-tight, leak-proof package that automatically folds to form a strong tray. It can be frozen, stored, reheated in a conventional oven, used as a serving dish, and easily disposed of. Similar packages are used in vending machines that dispense both cold and hot foods.

The technology of the folding carton is an exciting subject of study for both designers and sociologists. One of the central problems of human history is how to feed the world population. Packaging can provide part of the answer. Containers can be designed to preserve and ship staple and perishable foods anywhere in the world, and even to outer space.

Printing on Folding Cartons

Folding cartons are suitable for all printing processes but are particularly adapted to offset lithography and gravure. Often a transparent film is glued over windows before the blanks are folded. During the printing operations, windows and die cuts can be added to the cartons.

Special finishing and decorative treatments, such as embossing, varnishing, and texturing, can be applied before, during, and after printing. In addition, wax, sealers, and laminations can be added to protect the carton's contents from moisture or from sticking to the inner surfaces.

SET-UP PAPER BOXES

Set-up paper boxes, which are rigid, permanent, three-dimensional containers (Fig. 3-17), have been in use ever since the invention of paperboard. Especially attractive paper boxes were made in France at the end of the nineteenth century. They were designed to contain ladies' hats, fragrances, and fashion accessories. In Germany and England, toys and games were packed in cardboard boxes with colorful labels. In the United States, the first paperboard boxes were made in Boston in 1839 by Col. Andrew Dennison. The Dennison name is still associated with many kinds of paper products.

At first, cardboard was used to make set-up boxes, which were covered with colorful papers or fabrics and used by merchants to sell luxury products (Figs. 3-18 through 3-20). Set-up boxes are now available in many attractive styles and finishes. They are mainly used for prestigious products, cosmetics, fashion accessories, jewelry, and cameras (see Figs. 3-22 to 3-24 at the end of this section).

Construction of Paper Boxes

The basic materials used to construct set-up, or rigid, paper boxes are paperboard and covering materials (Fig. 3-21). In addition, plastics can be combined with them to make useful packages, including transparent plastic domes, windows, and thermoformed trays.

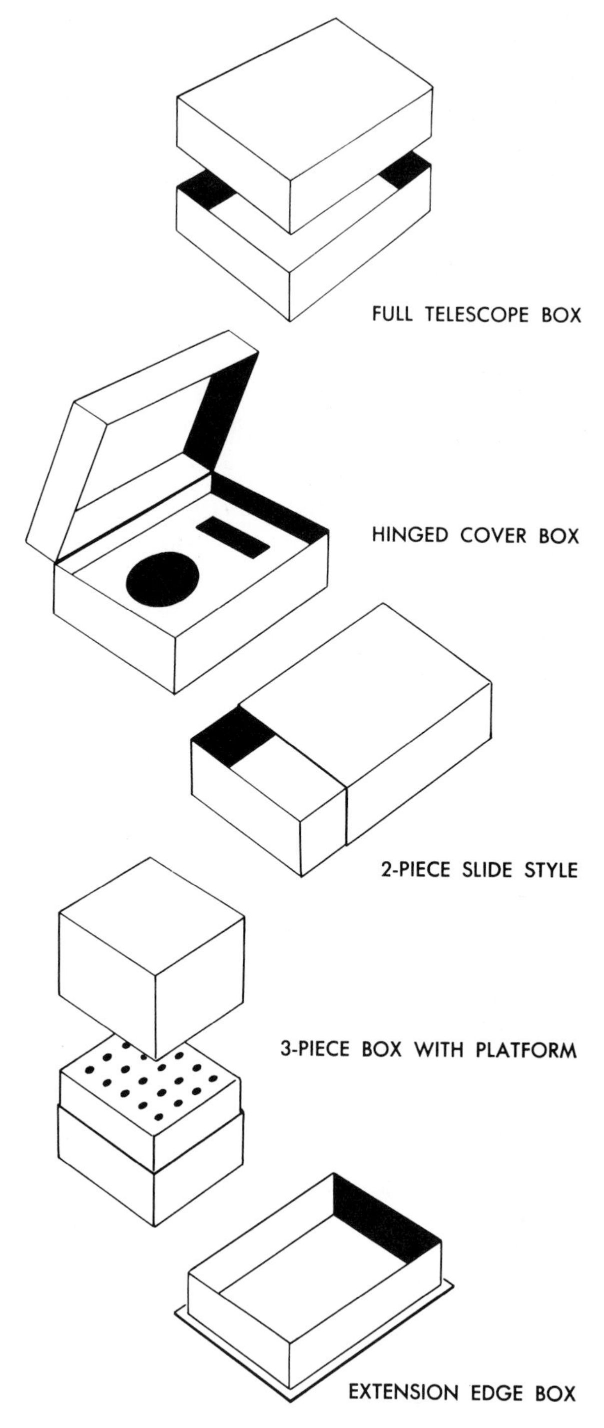

Figure 3-17. Some basic set-up boxes.

Figure 3-18. Soap sampler retail box, manufactured circa 1810. *(Photo courtesy of the National Paper Box Association.)*

Figure 3-19. Miniature hat box for doll's clothing, manufactured circa 1878. *(Photo courtesy of the National Paper Box Association.)*

Figure 3-20. Ladies' face powder compact, manufactured circa 1878, is a two-piece drawer type extension edge box. *(Photo courtesy of the National Paper Box Association.)*

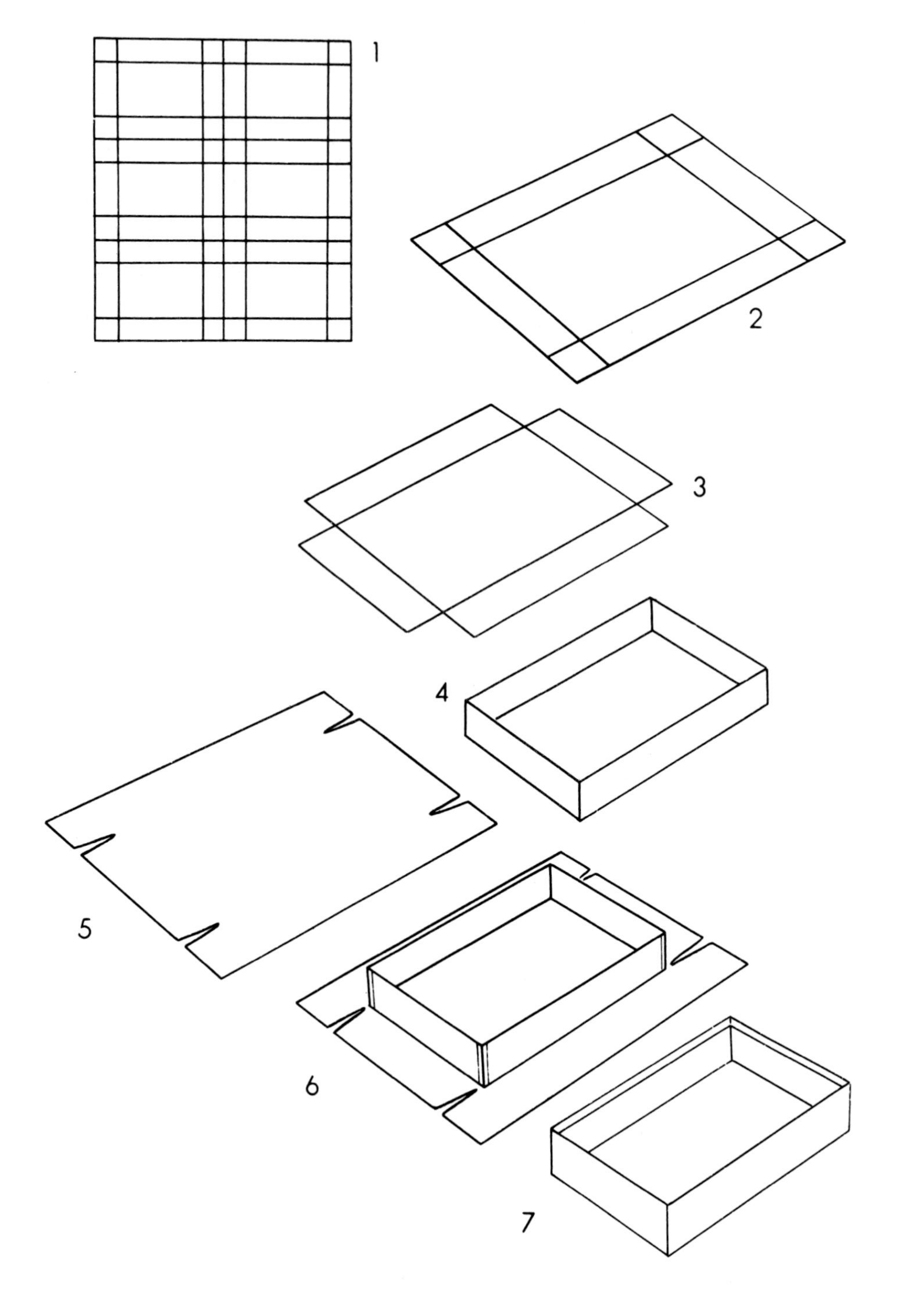

Figure 3-21. How to construct a set-up box.

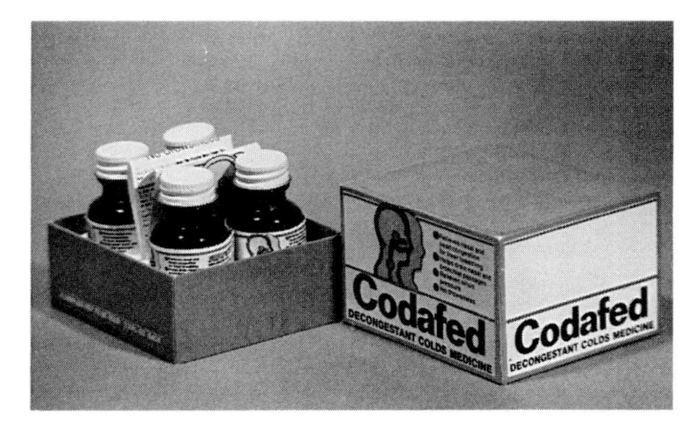

Figure 3-22. Set-up box for pharmaceutical product. Eliana Themistocleous, designer. *(Photo by Miles David Sebold.)*

Figure 3-23. Cosmetic set-up boxes. *Left to right:* Karen Whitlock, Eliana Themistocleous, and Susan Ceglio, designers. *(Photo by Miles David Sebold.)*

Figure 3-24. Set-up gift box. Susanne Bellezza, designer. *(Photo by Miles David Sebold.)*

Many types of papers and fabrics are used for box coverings (Figs. 3-22 through 3-24). These range from inexpensive wraps to embossed foils and lacquer spray-coated sheets. In addition, special designs can be developed and custom-printed. Accessory materials such as seals, tags, ribbons, and ties are used to enhance rigid boxes.

CORRUGATED CONTAINERS

Like most packaging materials, *corrugated board* (often called *fiberboard*) has a long and colorful history. When you were a child, corrugated boxes may have been your favorite playthings. As an adult, you probably pack your belongings in them whenever you move. Your TV set, stereo,

VCR, and other appliances were shipped in impressively designed corrugated boxes. It may surprise you to learn that this popular packaging medium, the workhorse of the industry, was originally part of an article of clothing. The nineteenth-century gentleman's tophat was fashioned with a sweatband of fluted paper, the precursor to corrugated board.

An American inventor, Albert L. Johnes, patented fluted paper for use in protective containers for bottles in storage and shipment. In 1874 another American, Oliver Long, invented a process for sandwiching the flutes between two sheets of paperboard. This innovation marked the beginning of a new industry—corrugated containers.

Today the corrugated container industry employs about 118,000 people in 1427 plants, producing 200 million boxes a year. It is a $10 billion industry, the largest in the paperboard packaging field. Its largest single market, representing more than one-third of the industry's output, is food packaging (U.S. Department of Commerce 1987 figures).

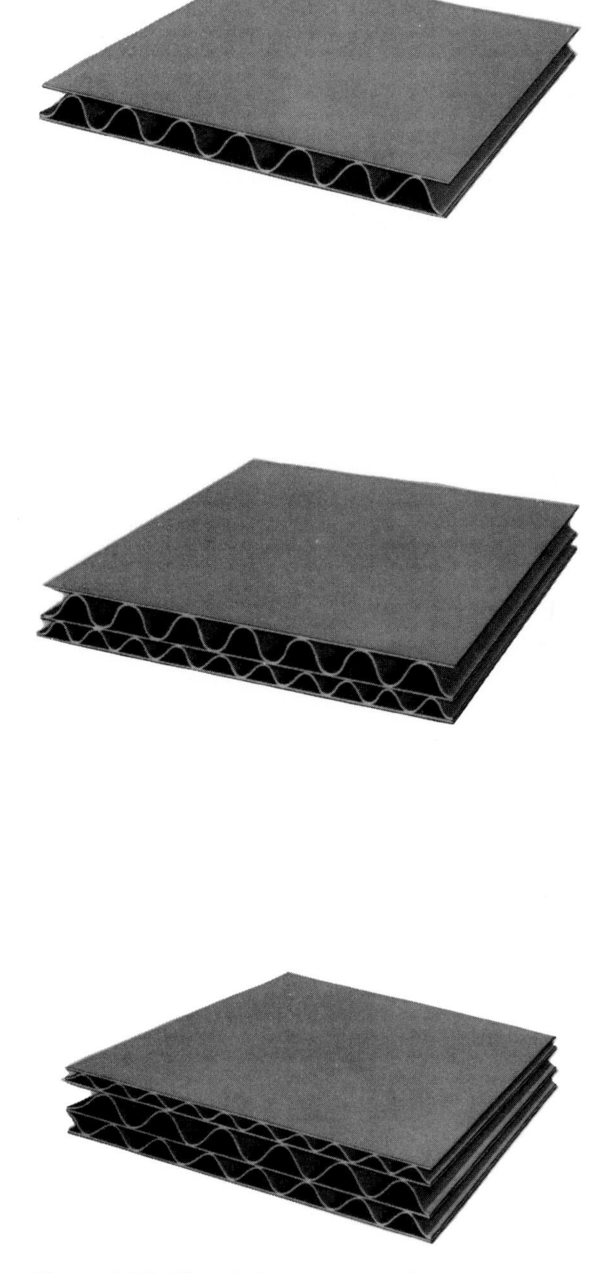

Figure 3-25. Fluted sheets are combined according to packaging requirements. *(Photo courtesy of the Fiber Box Association.)*

Construction of Corrugated Containers

Corrugated containers are constructed from a fluted sheet glued to one or more liners (Fig. 3-25). The structural characteristics of the corrugated medium are governed by four variables:

- The strength of the liners
- The strength of the corrugated medium

- The height and number of flutes per foot

- The type of walls (single, double, triple, etc.)

Four flute structures are available for corrugated containers (see Fig. 3-26):

- A-flute, in which wide spacing of flutes results in greater capacity to absorb shock.

- B-flute, which has a greater number of flutes per foot, providing maximum crush resistance.

- C-flute, which combines the properties of A and B flutes.

- E-flute, which is a very thin corrugated board and is perhaps most popular type for large, sturdy displays and packages.

Another important element of corrugated boxes is interior protection. A wide range of corrugated partitions, liners, pads, and other devices, including plastics (molded polystyrene foam) are used to provide inner reinforcement, cushioning, bracing, and shock absorption. The most commonly used closure techniques are stitching, stapling, gluing, and taping. Figures 3-27A and B show basic construction of corrugated boxes.

Although stock sizes of corrugated boxes are available, custom-made cartons are usually required for special jobs such as large quantities of products (10,000 to millions). Industrial or *master cartons*, which are shipping cartons with smaller cartons in them, are used for food, detergents, hardware, housewares, and so on.

Government and industry standards and regulations are designed to protect

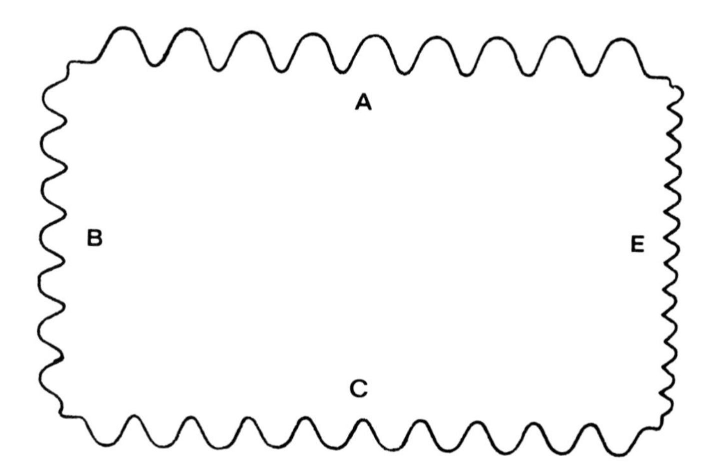

Figure 3-26. A corrugated flute gauge (actual size).

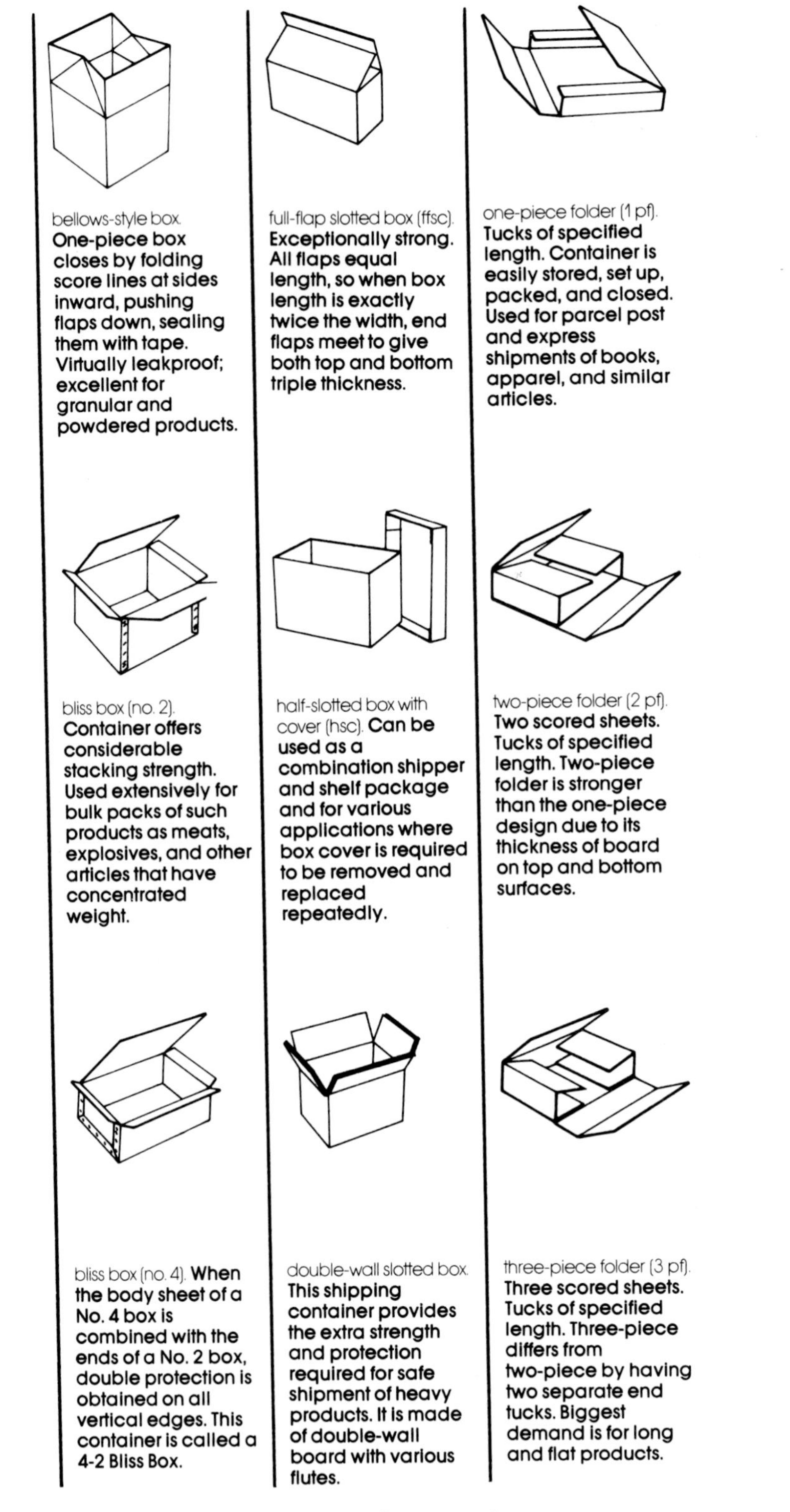

bellows-style box. One-piece box closes by folding score lines at sides inward, pushing flaps down, sealing them with tape. Virtually leakproof; excellent for granular and powdered products.
full-flap slotted box (ffsc). Exceptionally strong. All flaps equal length, so when box length is exactly twice the width, end flaps meet to give both top and bottom triple thickness.
one-piece folder (1 pf). Tucks of specified length. Container is easily stored, set up, packed, and closed. Used for parcel post and express shipments of books, apparel, and similar articles.
bliss box (no. 2). Container offers considerable stacking strength. Used extensively for bulk packs of such products as meats, explosives, and other articles that have concentrated weight.
half-slotted box with cover (hsc). Can be used as a combination shipper and shelf package and for various applications where box cover is required to be removed and replaced repeatedly.
two-piece folder (2 pf). Two scored sheets. Tucks of specified length. Two-piece folder is stronger than the one-piece design due to its thickness of board on top and bottom surfaces.
bliss box (no. 4). When the body sheet of a No. 4 box is combined with the ends of a No. 2 box, double protection is obtained on all vertical edges. This container is called a 4-2 Bliss Box.
double-wall slotted box. This shipping container provides the extra strength and protection required for safe shipment of heavy products. It is made of double-wall board with various flutes.
three-piece folder (3 pf). Three scored sheets. Tucks of specified length. Three-piece differs from two-piece by having two separate end tucks. Biggest demand is for long and flat products.

Figure 3-27A. Basic corrugated construction.

single- and double-lined slide boxes. **Single lined used as interior container and for parcel post, express, freight shipments. Double lined is three piece and provides double thickness all sides.**

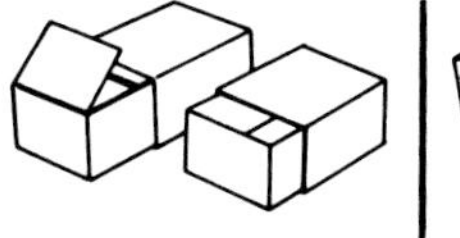

triple- and double-slide boxes. **Triple slide (left) has two thicknesses of double-faced board all sides; double slide has two thicknesses on two sides. Collapsible; ideal for mail and express.**

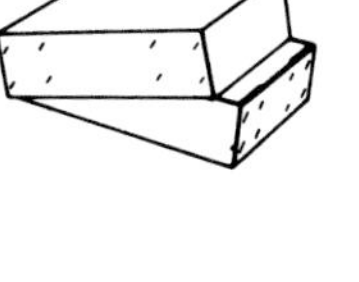

telescope design box. **Extra thickness of corrugated board in side and end walls of this two-piece container affords exceptional stacking strength and overall protection to the contents of container.**

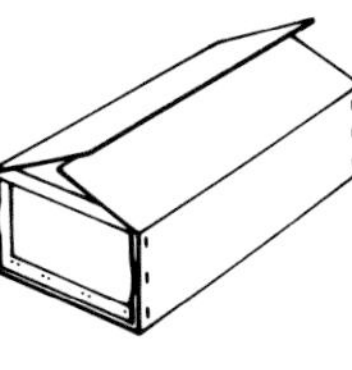

recessed-end box. **Three-piece box has a body sheet and two flanged end pieces. By varying body-sheet length, box size can change to fit many products of same girth but different lengths.**

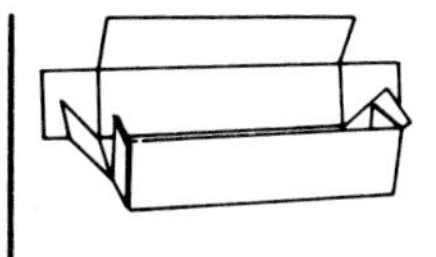

five-panel folder. **Use for canes, umbrellas, similar long, slim items. Each end has a minimum of three thicknesses, providing strength where it is most needed. Container is shipped flat to user.**

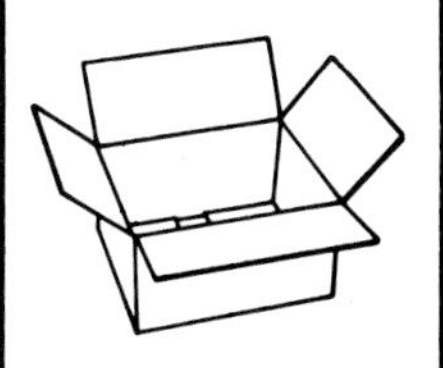

regular slotted box (RSC). **Top and bottom flaps are equal length; folded inner flaps meet only if box is square. Securely sealed with adhesive, gummed tape, or metal stitches as desired.**

half-slotted box with half-slotted partial cover (pths). **Two-piece box, both sections slotted style. Double thickness of corrugated provides great resistance to bulging and buckling.**

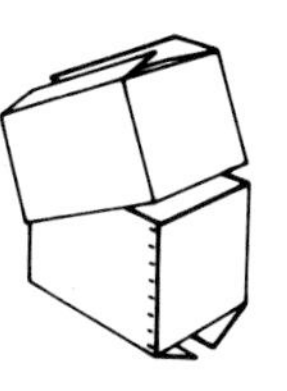

full-telescope, half-slotted box (fths). **Full-depth cover, two-piece box. Both sections of slotted style. Full-cover top renders maximum product protection and superior stacking strength.**

center special slotted box (CSSC). **Construction gives double-flap thicknesses top and bottom. One or both side flaps are shorter than end flaps, so all flaps meet for double top and bottom.**

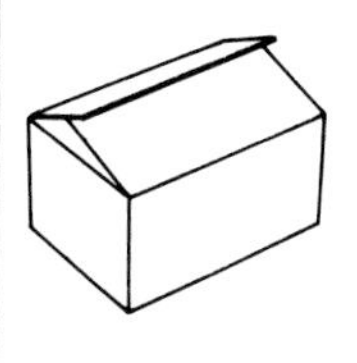

overlap slotted box (osc). **Efficient when products packed for shipment require sealing with metal staples, stitches, straps. Side flaps partly overlap for added rigidity at both top and bottom.**

double-thickness score-line box (Box With Cover). **Another design meeting requirements for double thickness score-line box under Railroad Shipping Rule 41. Box carries heavy loads despite rough handling.**

double-cover box. **Popular with manufacturers of articles that cannot be readily packaged in standard containers. In large sizes, this box is often used as a unitized or palletized load.**

double-thickness score-line box (Conventional Slotted Style Box). **For high-density products (screws, nuts, washers) in weights to 300 pounds. Container is fast replacing wooden nail kegs.**

design box with cover. **Space-saving, stapled box with double end flaps and lid. Especially easy to pack, the design box is used for shipment of cut flowers, wreaths, and similar products.**

interlocking double-cover box (ic). **Flanges on covers interlock with flanges on tube. Three-piece box for items under Railroad Shipping Rule 41. Greatest use in packing heavy appliances.**

Figure 3-27B. Basic corrugated construction.

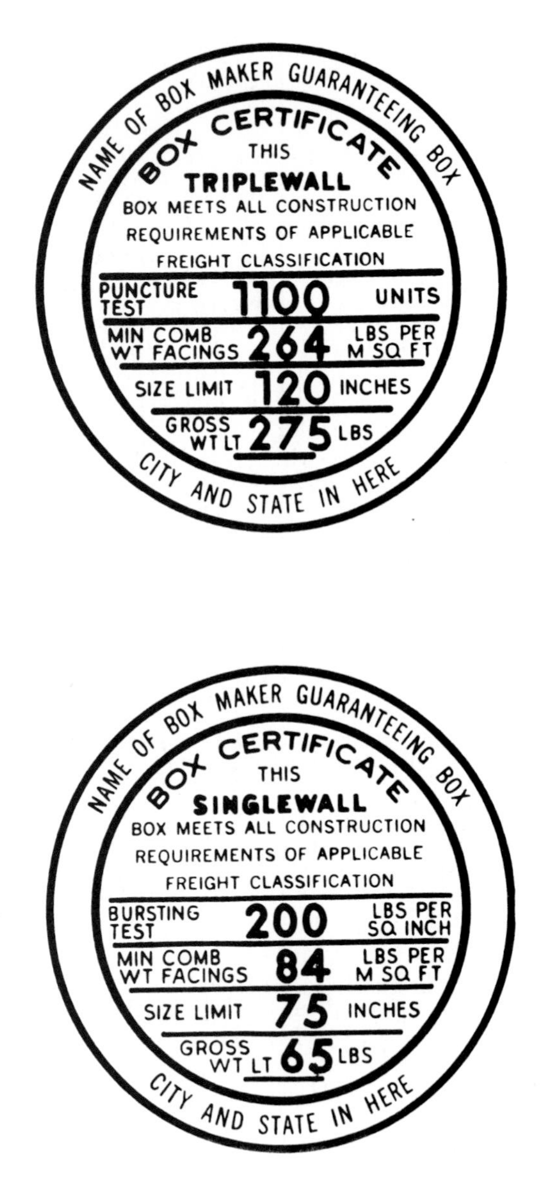

Figure 3-28. Weight, paper content, and puncture and bursting test certificates must be displayed on all corrugated containers.

users of cartons. There are laws pertaining to shipment method, such as rail, air freight, truck, and regular parcel post. All corrugated materials and cartons must be certified by the manufacturer. Weight, paper content, and puncture and bursting test certificates must be displayed on all corrugated boxes (Fig. 3-28).

The tests for the materials involve subjecting them to the same conditions to which a product is subjected in the course of normal handling; these include drops, jolts, shocks, and vibrations. These tests are designed to select the right box for a product, as well as the right master carton in which to ship the boxed product, without costly packaging, overpackaging, or underpackaging.

A significant trend in corrugated technology is impregnating and coating the corrugated board with waxes and plastics. The moisture-resistant coating permits reuse of cartons to ship products such as fruits, vegetables, and other products that were previously shipped in expensive wooden crates and barrels.

Printing on Corrugated Materials

The most important advance in printing on corrugated materials is *flexographic printing*, which uses quick-drying inks and high-speed presses. Another significant trend is toward greater use of colors on clay-coated white liner board. The use of preprinted liners and full-color lithographed, laminated labels turns a common cardboard box into an elaborately printed package (Fig. 3-29).

Figure 3-29. Corrugated "E" flute, full-color packages. Susan Fanuzzi, designer. *(Photo by Miles David Sebold.)*

POINT-OF-PURCHASE DISPLAYS

The expansion of the self-service stores and the change in consumer buying habits have both contributed to the development of point-of-purchase (POP) displays. The POP display and its variations have become an effective sales aid for retailers.

Brief History of Point-of-Purchase Advertising

Point-of-purchase (POP) advertising may be one of the oldest forms of communication. In ancient Rome, merchants often displayed the symbols of their trade to attract customers. Archaeologists excavating

Pompeii and Herculaneum found remnants of metal trade signs forged by blacksmiths. Some traces of hand-painted signs were also found at the entrances to shops.

During the Middle Ages, merchants advertised their wares with models of keys, tools, swords, utensils, pottery, and other products. Since most customers were illiterate, it was necessary to use clearly understood symbols. During the Renaissance, illustrated signs began to appear throughout Europe (see Figs. 3-30A-C).

Figure 3-30B. Carved European trade sign (seventeenth century).

Figure 3-30A. Carved European trade sign (seventeenth century).

Figure 3-30C. Iron European trade sign (eighteenth century).

These constitute the first known instances of collaboration between artists and tradesmen. Some of these signs and symbols are still in use today. A symbol that is still used to indicate a pawn shop began as the three golden balls of the Medici (Fig. 3-31), who were originally bankers and money lenders. The striped barber pole originally represented the establishments of barber-surgeons in the days when the two professions were indistinguishable. These popular symbols have survived for hundreds of years.

Figure 3-31. Pawn shop sign. *(Photo by author.)*

In the United States in the nineteenth century, the most popular point-of-purchase display was the cigar store Indian (Fig. 3-32). These 6-foot wooden statues were hand-carved from white pine, usually by the same people who carved figureheads for ships. They were painted realistically and placed at the doorways of cigar stores. Collectors estimate that between 1850 and 1890 about 20,000 Indians were carved. Today they are valuable collectors' items.

In the early nineteenth century, merchants created their own sales aids with the help of sign painters. And in pioneer days western saloon keepers and general-

Figure 3-32. The cigar store Indian, an early POP display.

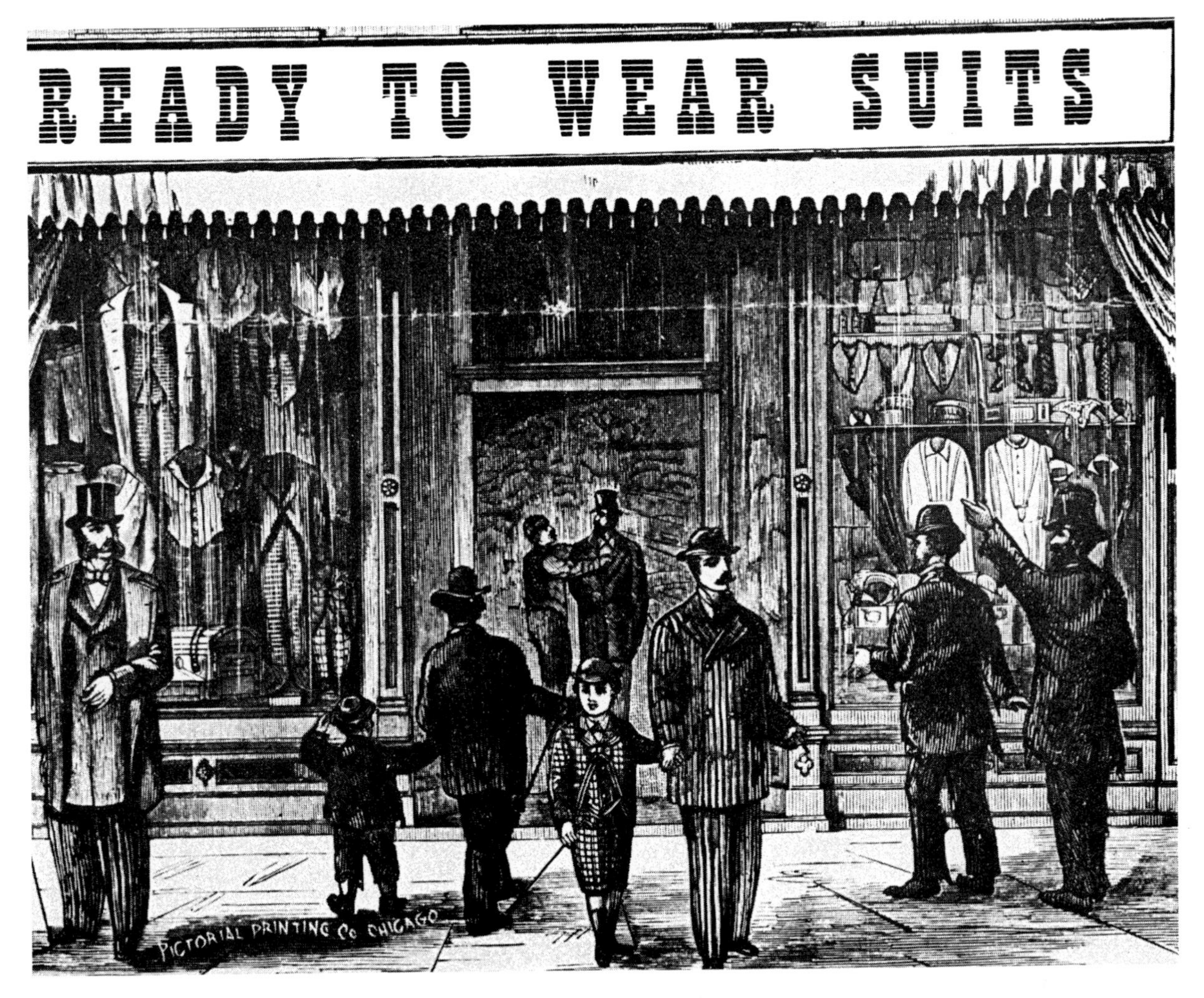

Figure 3-33. Store front and windows circa 1870.

store merchants were constant sources of original signage, mostly based on contemporary decorative typefaces (Fig. 3-33).

At about the same time, brand names began to appear on barrels, crates, tins, jars, and calendars (Figs. 3-34A and B). Manufacturers began to supply merchants with store fixtures marked with their signs (Fig. 3-35). It was considered quite prestigious to display these items. When glass display windows began to appear about 1840, consumers had the opportunity to see and admire merchandise that was cleverly and often profusely displayed in store windows (Fig. 3-36). In

Figure 3-34A. Soap wrap circa 1880.

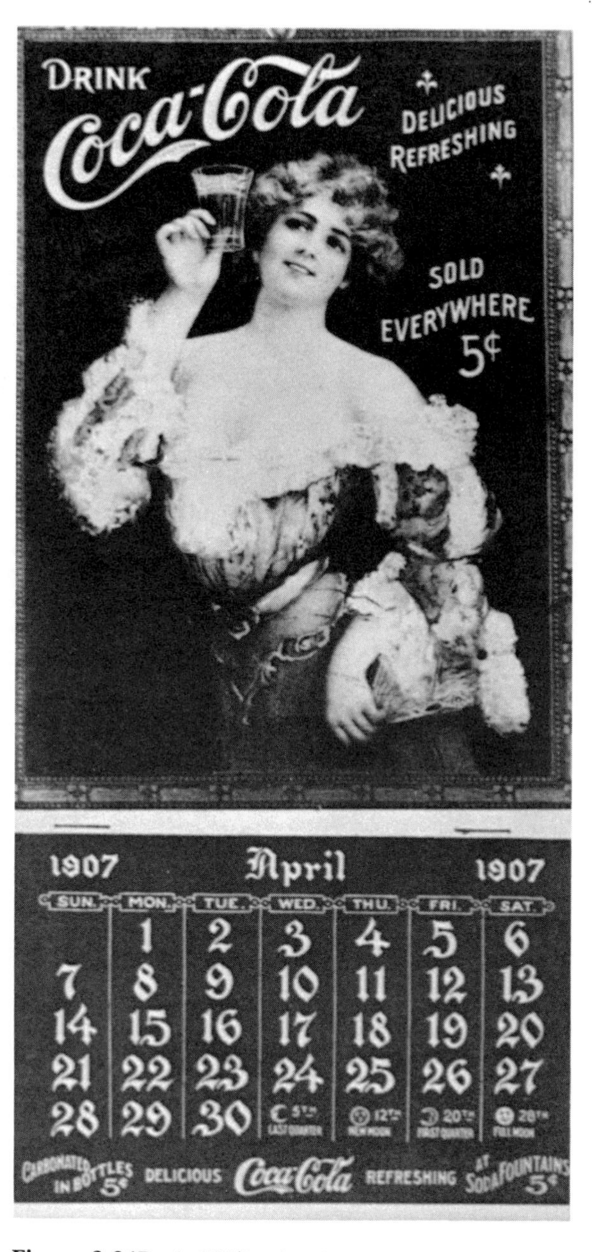

Figure 3-34B. A 1907 calendar premium.

Figure 3-35. A POP fixture from the nineteenth century.

Figure 3-36. A turn-of-the-century store window.

the countryside, metal or painted signs on barns and blacksmith shops advertised products and services.

Three-dimensional window displays did not appear until the early twentieth century. At first, store windows had an overstuffed look, since merchants displayed large quantities of merchandise in order to demonstrate the great variety of goods available in their stores. Then about 1910 die-cut three-dimensional displays began to appear (Fig. 3-37), and by 1922 the neon sign was widely available.

As merchandising, retailing, and selling techniques have become more and more sophisticated, so have the techniques of the POP display. Today's POP display is one of the most important selling aids in retailing and is crucial to the package designer's efforts.

Figure 3-37. A die-cut display card circa 1910.

Types of POP Displays

There are several types of POP displays, each serving a specific merchandising function. The major categories are

- Display merchandisers and shippers
- Permanent displays
- Window and showcase displays
- Posters
- Signage
- Commercial vehicles

Each category has several variations, depending on the purpose, location, and merchandising function of the displays.

Display Merchandisers

Retail establishments are busy places. A large percentage of the selling therefore takes place through self-service, whereby consumers choose merchandise themselves rather than having it brought to them by a salesperson. This makes impulse buying possible.

The types of displays used for impulse buying are *display merchandisers*, sometimes called *promotional displays* because they are designed to be used only for the duration of a particular sales promotion. These displays are strategically placed within the store, often near the checkout counter where the customer has time to look over the promotion while waiting in line. Structurally, the display merchandisers are designed for easy assembly.

A variation of the display merchandiser is the *counter display*, often called a *shipper* (Figs. 3-38 through 3-40). Generally a small display, it is popular with stores that sell cosmetics, health and beauty aids, and other small items. An interesting variation of the counter display is the *gravity-fed display*, which dispenses batteries, films, shoe polish, cosmetics, pouches, and similar objects (Fig. 3-41).

Another variation of the display merchandiser is the *floor stand*, a large structural display used mostly in supermarkets and liquor stores. It is often animated

Figure 3-38. A promotional counter display. *(Photo courtesy of MEM Company, Inc.)*

Figure 3-39. Counter display shippers. *(Photo courtesy of Warner/Lauren Ltd. Cosmetics.)*

Figure 3-40. Counter display with testers. *(Photo courtesy of Warner/Lauren Ltd. Cosmetics.)*

Figure 3-41. Gravity-fed display. *(Photo courtesy of Warner/Lauren Ltd. Cosmetics.)*

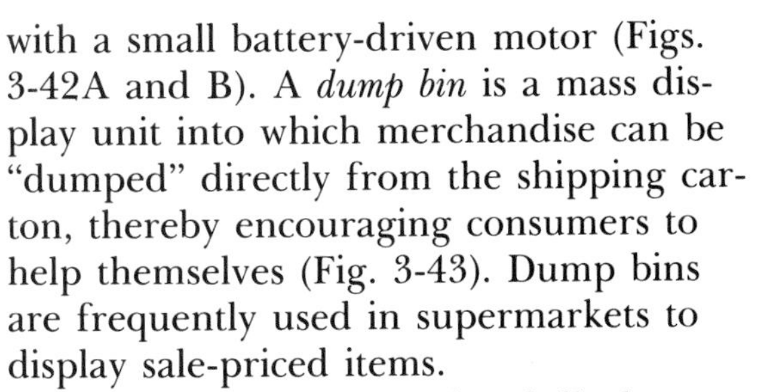

with a small battery-driven motor (Figs. 3-42A and B). A *dump bin* is a mass display unit into which merchandise can be "dumped" directly from the shipping carton, thereby encouraging consumers to help themselves (Fig. 3-43). Dump bins are frequently used in supermarkets to display sale-priced items.

Other types of promotional displays are used both inside of stores and outdoors, such as at gas stations (Fig. 3-44). They include mobiles, flags, banners, spinners, and devices that move with or without electric power.

Another POP device used to stimulate sales is the "send away" *premium display*,

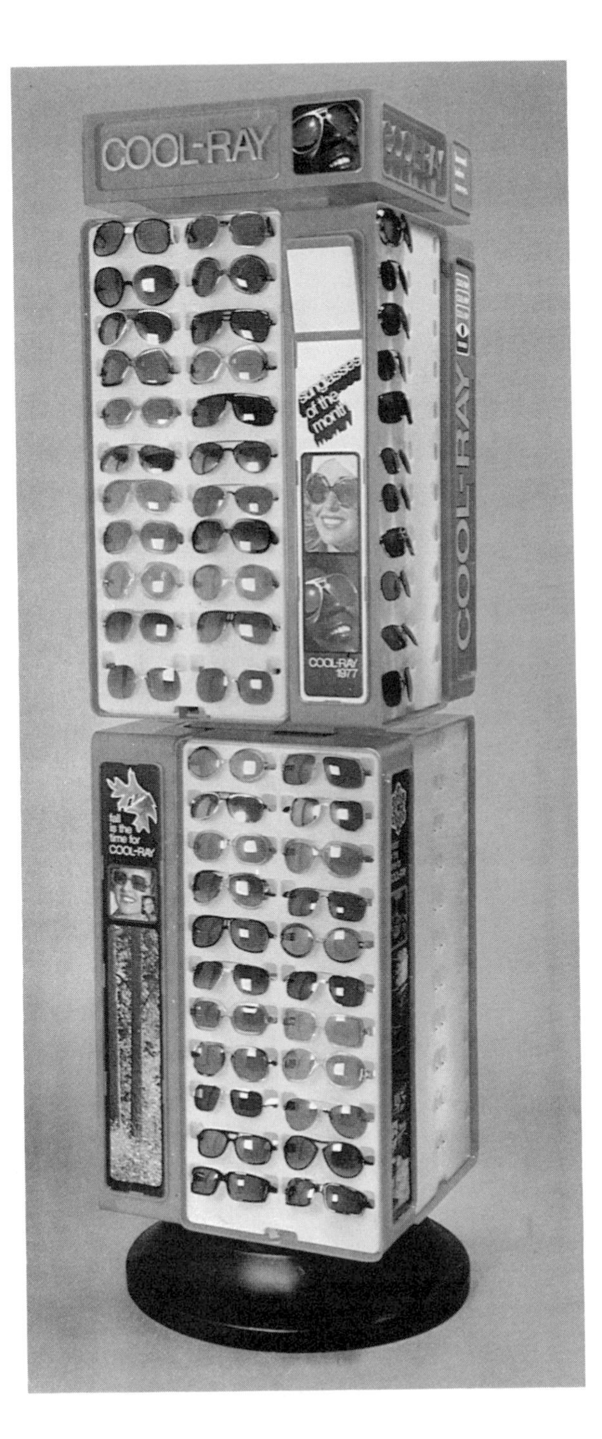

Figure 3-42A. Rotating floor stand. *(Photo courtesy of Cool-Ray Personal Care Brands, division of Warner-Lambert Company.)*

Figure 3-42B. Floor stand. *(Photo courtesy of Hallmark Cards, Inc.)*

Figure 3-43. Dump bin wire display. *(Photo courtesy of Brand's Fine Flavors.)*

Figure 3-44. Outdoor metal sign. *(Photo courtesy of Mobile Oil Corporation.)*

shown in Figure 3-45. This more elaborate type of stand is made to entice dealers into prominently displaying a product.

In addition, there are a number of promotional display graphic devices, such as shelf extenders, danglers, and die-cut signs (Fig. 3-46), designed to catch the attention of the consumer.

Figure 3-45. A dealer incentive premium display. *(Photo courtesy of National Distiller Products Company.)*

Figure 3-46. Shelf extenders, danglers, and signs. *(Photo courtesy of the Great Atlantic & Pacific Tea Company, Inc.)*

Permanent Displays

Sometimes temporary promotional displays are not suitable for the store or product. In such cases, permanent displays may be used (Fig. 3-47). *Permanent displays* are made of durable materials such as wood, plastics, and wire and tend to be elaborate and expensive. They are closely associated with cosmetics. Their functions include selling, demonstrating, and helping consumers sample products (e.g., fragrance testers and shade charts).

Window and Showcase Displays

The most valuable type of POP display is probably the *window display*, since they attract customers from the outside. Window displays range from a simple poster to

Figure 3-47. Permanent, gravity-fed display. *(Photo courtesy of Helena Rubinstein, Inc.)*

the intricate displays seen in the windows of travel agencies and liquor stores. Since window displays are created by the store's display department, technically they are not a POP display.

Posters

The *poster* is the oldest graphic promotional device. Some of the most prominent artists of this century, including Toulouse-Lautrec and Picasso, created posters for theaters, ballet companies, art galleries, and political movements. The art of poster design is highly expressive, making posters powerful social and political weapons as well as works of art (Figs. 3-48 through 3-51).

Figure 3-48. Henri Toulouse-Lautrec color lithograph from the author's collection.

Figure 3-49. James Montgomery Flagg's World War I recruiting poster.

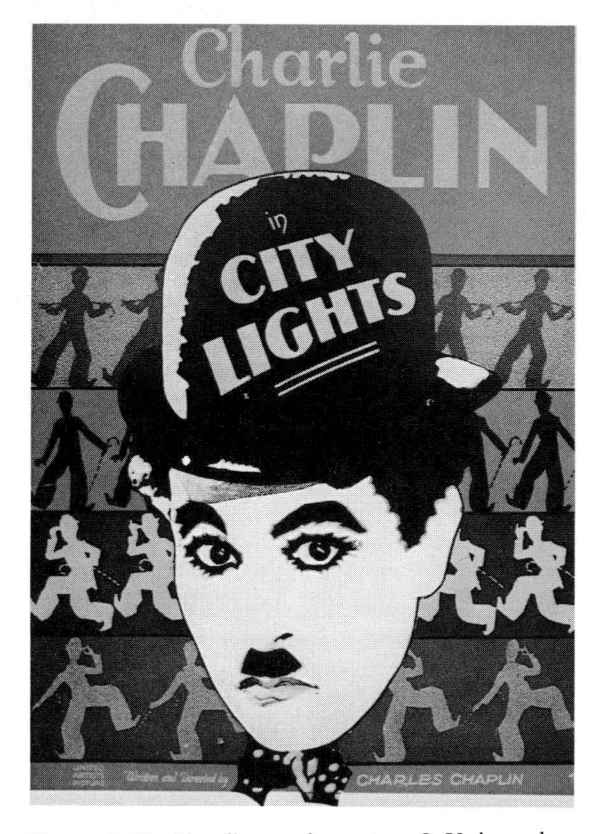

Figure 3-50. Chaplin movie poster. © Universal Pictures. *(Photo courtesy of MCA, a division of MCA Communications, Inc.)*

Figure 3-51. Poster by Seymour Chwast, designer.

Signage

The POP graphic displays that identify companies and products and convey messages about them is called *signage*. This type of POP is frequently used by fast-food outlets and gas stations, where a total environment is created with signs. Automobile dealerships, appliance showrooms, international fairs, and the Olym-

Figure 3-52A. Contemporary trade sign. George Wybenga, designer.

Figure 3-52B. Contemporary trade sign. George Wybenga, designer.

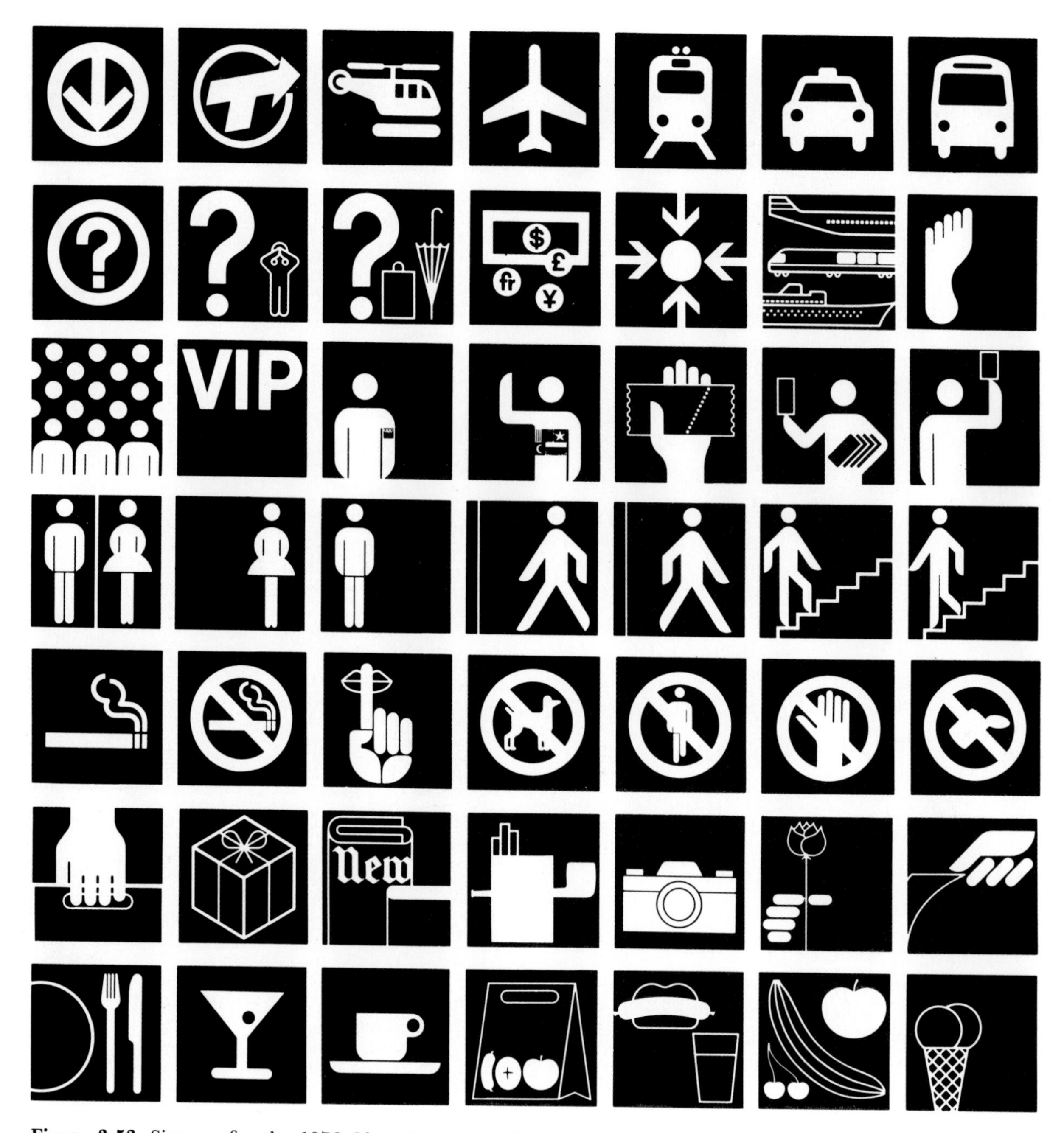

Figure 3-53. Signage for the 1976 Olympic Games, Montreal, Canada. Hunter, Straker, Templeton, Ltd., Toronto, designers. *(Photo courtesy of the Bicultural Information Committee of Public Works, Canada.)*

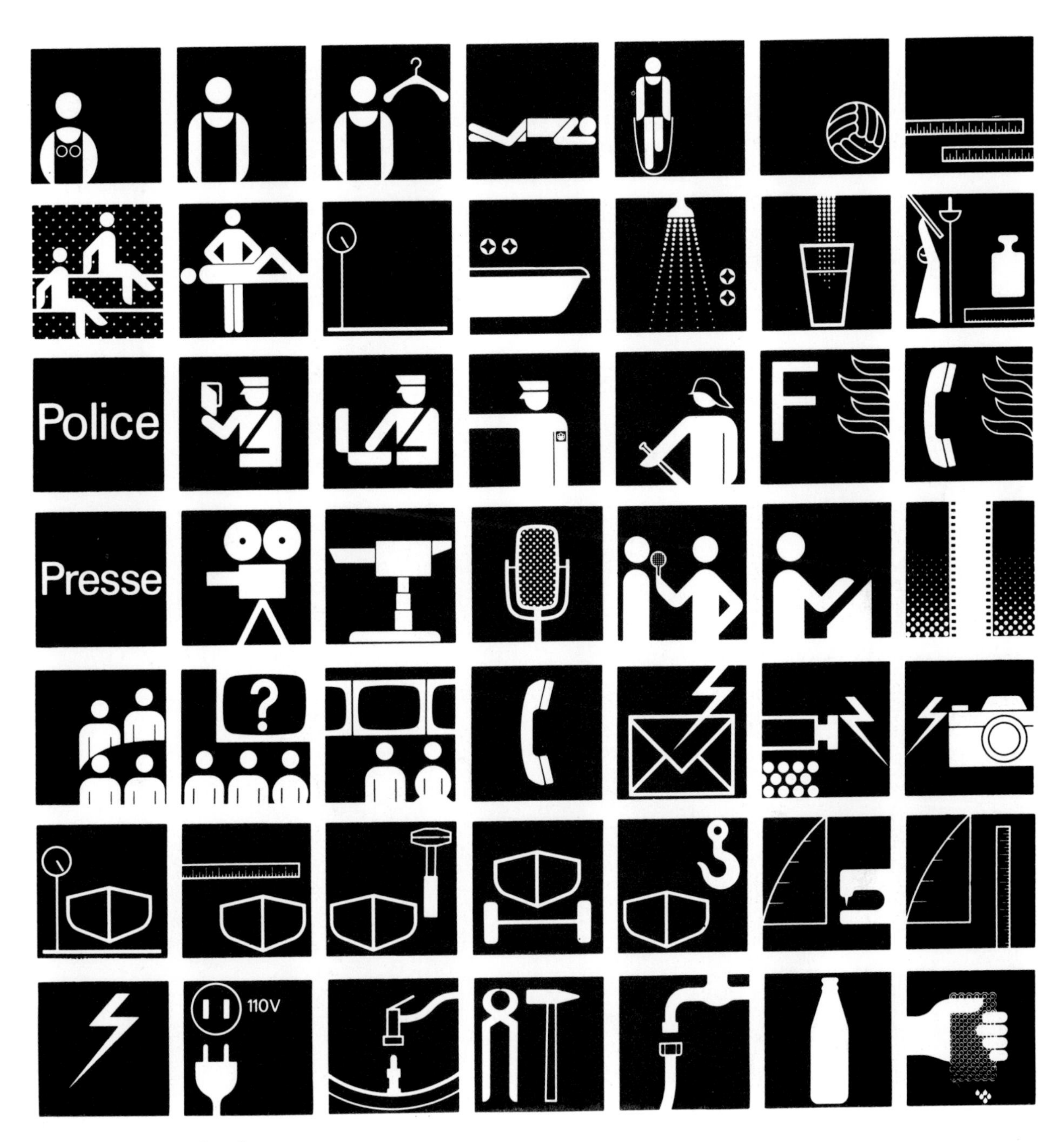

Figure 3-53. *(continued)*.

pic Games are other examples of places in which signage can be used effectively (Figs. 3-52A and B, and 3-53).

Soft-drink, beer, and camera manufacturers distribute wall plaques, clocks, and indoor and outdoor signs—often using neon—to bars and restaurants. These POP "signs" are an advertising device as well as a service to consumers.

Commercial Vehicles

The *commercial vehicle* is another valuable selling device. They include circus wagons, ice cream carts, trucks, and airplanes, all bearing the name and logo of the company. Commercial vehicles can be thought of as moving billboards (Figs. 3-54 through 3-56).

Figure 3-54. Early signage on a delivery truck. *(Photo courtesy of General Foods Corporation.)*

Figure 3-55. Goodyear blimp. *(Photo courtesy of the Goodyear Tire and Rubber Company.)*

Figure 3-56. Corporate identity logo. *(Photo courtesy of Federal Express Corporation.)*

Designing a POP Display

The POP designer is not just an artist who can use color, art, and typography. He or she must have a thorough understanding of the problems faced by various types of retail establishments. To create an effective POP display, the designer must know marketing and advertising procedures, the layout of the retail establishment, as well as design techniques.

As consumers, POP designers are probably familiar with the layouts of supermarkets, drugstores, and package stores; but as designers they must also understand the techniques used to attract customers and sell the product. It may seem a simple matter to design a display on a drawing board, but it is quite another one to build and produce it—and remain within the required budget. This is no small undertaking.

As discussed in Chapter 2, the first step in learning about retailing is to visit a retail store. Four aspects of the store should be studied: the physical layout, the flow of customers, the placement of the various types of displays, and demographic and ethnic preferences for different merchandise (obtained from professional studies by state and federal government and private companies). Valuable information can also be obtained from the store manager, who is usually willing to listen to the designer and suggest improvements and changes in proposed designs.

The designer also needs to study the manufacturing methods and packaging systems used to create the products to be displayed. Such knowledge provides answers to questions such as the following:

How should the product be packed and shipped?

What problems are likely to be encountered in getting the product into the display?

How should the display be shipped to the retailer?

These and related questions must be answered as part of the process of designing an effective POP display.

Advertising, marketing, engineering, and production personnel also need to be consulted throughout the design process. Like many other design projects, the design of a POP display requires teamwork.

As for any packaging concept, the concept for a POP display is first presented as a marker sketch. Once the display concept has been approved, a full-scale model is constructed. Since most cardboard displays are shipped unassembled, it is advisable to work out the assembly system for the display *before* building the model. Include a how-to assembly instruction sheet.

The most important consideration affecting a POP design is the budget. Many types of displays are possible, ranging from small, inexpensive shipper merchandisers made of E-flute corrugated cardboard to costly permanent displays made of plastic, metal, and wood. The advent of flexographic printing on corrugated E-flute has made it possible to produce attractive full-color displays at reasonable cost. Other types of corrugated E-flute displays, merchandisers, counter displays, floor stands, and dump bins can be designed with effective graphic applications. Their cost depends on the type of display, which can range from simple screen or flexographic printing to complex photographic effects in which a full-color lithography serves as the merchandiser or promotional display. This versatile POP display can be used to sell books, toys, cosmetics, housewares, and many other kinds of merchandise.

PROJECTS

1. Make a rough sketch of a small counter display or dump bin for a specific item of merchandise (e.g., books, candies, cosmetics). Work out your structural problems in a sketch. Construct your display from white corrugated E-flute or cardboard. For the dump bin, build a scale model. (The average dump display is about 6–7 feet tall.)

2. Birdseed usually comes in a plastic bag. Design and build a box that can be turned into a bird feeder. Use impregnated B- or E-flute corrugated. Place the plastic bag containing the birdseed inside the box. Leave an opening for puncturing the bag and devise a small protruding platform for the seed.

3. To better understand the "anatomy" of the folding carton (tray or tube), it is helpful to construct several "blanks" with the help of the box patterns shown on these pages. Construct your own variations on these patterns. Use a strong 2- or 3-ply Bristol board. For an adhesive, use white glue only.

SUGGESTED READINGS

Griffin, Roger C., Jr., and Stanley Sacharow. *Principles of Package Development.* Westport, CT: AVI, 1972.

Roth, Lászlo. *Display Design.* Englewood Cliffs, NJ: Prentice Hall, 1983.

4

THE AGE OF PLASTICS

Natural Plastics / Development of Synthetic Plastics / Chemistry of Plastics / Classification of Polymers / Techniques Used to Mold and Shape Plastics / Designing with Plastics / Other Important Uses of Plastics

THE word "plastic" comes from the Greek word meaning "fit of molding." For many years plastic was used to describe materials that could be easily molded or changed in shape by being pressed into simple molds or dies. In time, an industry developed in which objects were made from substances created in chemical laboratories. The objects were called *plastics* because many of them were molded. When we think of plastics today, we usually think of man-made materials or substances.

We are, in fact, living in the age of plastics. A quick look around any home or office will reveal hundreds of plastic products and materials. We sit on plastic-covered chairs, eat from plastic plates, drink from disposable plastic cups. Food is brought home from the supermarket in airtight plastic bags or trays and cooked in frying pans with nonstick plastic coatings. Some of our clothing is made of plastic fibers. We have plastic in and on our teeth; we wear plastic glasses or contact lenses. Modern medical procedures use plastic devices to replace organs and limbs. Building materials, vehicles, fabrics, and paints are created from plastics.

Plastics can be hard or soft. Some are flexible, some transparent. Plastics come in many forms—liquid, solid, fibers, foams, films—and can be molded in a vast range of shapes and colors. Plastics are, in short, among the most useful and versatile materials available.

NATURAL PLASTICS

When Christopher Columbus visited the Americas, he is reported to have seen Indians playing with balls made from a kind of gum (Fig. 4-1). Other explorers learned that the Indians made waterproof "shoes" by dipping their feet in the same gum and allowing it to dry. The gum was *latex*, which is produced by rubber trees.

Figure 4-1. Aztec Indians playing with rubber ball. Detail from the Aztec "Telleriano Codex."

The name "rubber" was given to this substance in 1770 by the English chemist Joseph Priestly, who discovered that in addition to its other useful qualities it rubbed out marks made from pencil.

Rubber

By 1820 a small rubber industry had developed in England. In 1824, Charles Macintosh began manufacturing waterproof raincoats by spreading a thin layer of rubber over fabric. In 1839, the American inventor Charles Goodyear discovered the process of *vulcanization*, which strengthened rubber and made it more resistant to heat and cold.

Rubber originally came from rubber trees in Brazil. But the demand for rubber soon began to outstrip the supply available from Brazil. Therefore, in 1876 an English botanist, Henry Wickham, took several thousand rubber tree seedlings to Ceylon and Malaysia, where they flourished. Eventually they developed into the giant rubber plantations of Southeast Asia, the source of most natural rubber used today.

Natural Plastic Substances

Many other plastic substances are found in nature. *Amber*, a fossil resin, is a gum exuded by certain ancient trees. It is used to create jewelry. *Rosin* is a residue from pine trees that is used in turpentine and waxes, to treat violin bows, and to make soaps and varnishes. *Lac* is a substance derived from the bodies of scale insects. It is used in preparing *shellac*. *Gutta-percha*, a rubberlike substance, is exuded by rubber trees in Malaysia.

The ancient Egyptians, Greeks, and Romans used *bitumen* to seal documents, waterproof ships, and line water tanks. Bitumen takes many forms, including pitch, tar, and asphalt. It is found throughout the world. One well-known source, Pitch Lake in Trinidad, was discovered in 1595 by the English explorer Sir Walter Raleigh.

The plastic properties of *ivory* and *horn* have been known for some time. Both of these "plastics," or actually animal tusks, have been used for jewelry and carvings for centuries.

Still other natural plastics include *casein* (a derivative of milk), which is used for making paint, and *plaster of paris* and *cement*.

DEVELOPMENT OF SYNTHETIC PLASTICS

The history of synthetic plastics dates from the early nineteenth century, when great strides were made in all branches of science. Among other research efforts, chemists began searching for substitutes for ivory, the principal source of material for piano keys and billiard balls.

Around 1850 plastic molders mixed gutta-percha with other natural resinous binders and, using complex steel molds, formed them into small novelties such as frames, toys, combs, and toothbrush handles. In 1862 the English chemist Alexander Parkes found that if cotton was treated with acid and mixed with camphor, a hornlike substance was formed. He called this substance *Parkesine* and exhibited it at the Great Exposition of 1862. Parkes was not a businessman, however, and his efforts to market products made from Parkesine failed.

In 1868 two New York manufacturers, Phelan and Collender, offered a prize of $10,000 to anyone who could develop a practical substitute for ivory to be used in making billiard balls. A Newark printer, John Wesley Hyatt, and his brother Isaiah took the challenge. They started with the compound *cellulose*, which is found in plants and is ordinarily produced from cotton. They treated the cellulose with sulfuric acid and nitric acid. When only a part of cellulose was changed by the acid, a compound called *pyroxylin* was formed. The Hyatt brothers combined pyroxylin with camphor to obtain what we call *celluloid*.

Celluloid quickly came into widespread use as a substitute for hard rubber. Collars, toys, dolls, novelties, and motion picture film were produced from this early synthetic plastic. Celluloid had several defects, however. Chief among them was that it was dangerously flammable. The plastics industry therefore began searching for ways in which to produce less dangerous, yet equally useful, cellulose compounds.

The plastics industry made a major breakthrough in 1909, when Leo H. Baekeland announced his discovery of phenol-formaldehyde resins. These were the first thermosetting (heat-hardening) plastics and were called *bakelite*. They are widely used in practically every industry today.

Cellulose acetate was discovered in 1865, but it wasn't used commercially until 1912. In 1922 the Swiss chemist Jacques Brandenburger developed a machine to produce a material that he called *cellophane*. This transparent film soon found many uses.

Although some of the scientific discoveries that contributed to the development of the plastics industry occurred in the middle of the nineteenth century (e.g., polystyrene, polyvinylidene chloride, polyester, and rubber hydrochloride), the most significant practical developments did not take place until the 1930s. By 1950 virtually all the plastics that are familiar to us today were being produced commercially.

Today plastics are synthesized from crude oil, coal, natural gas, air, water, limestone, salt, cotton, soybeans, corn, and substances obtained from trees. In recent years manufacturers have developed new plastics such as ABS, PET, Lexan, resins, and "high-performance" plastics, which are used for bottles and containers since they can be recycled (i.e., returned and refilled). In addition, chemical engineers can build many desirable features into plastics such as flexibility, strength, transparency, and biodegradability.

CHEMISTRY OF PLASTICS

In studying the basic chemistry of plastics, the first step is to become familiar with some chemical terms. All matter is made of minute particles called *atoms*.

The combining power of an atom to join with other atoms is called its *valence* from the Latin word meaning "power."

An *element* is a basic structure that is made up of only one kind of atom; for example, the element hydrogen contains only hydrogen atoms. A *compound* is a substance formed by the chemical union of two or more elements or ingredients. The compound then has properties that are different from each of the individual elements. Carbon dioxide, for example, is the union of carbon and oxygen. The compound carbon dioxide has different properties from the individual elements carbon and oxygen. A *molecule* is the smallest particle of an element or compound. It retains its chemical identity; that is, it does not change chemically.

Figure 4-2. Plastic resins (polymers).

Chemists use the term *polymer* (or *resin*) to refer to the chemical compounds or mixtures used to make plastics and plastic products (Fig. 4-2). *Poly* comes from the Greek word for "many," and *mer* means "part." Polymers are commercially produced as liquids, pastes, granules, powders, and flakes.

A polymer is a large chemical molecule with many repeating parts, or links. In this respect it can be compared to a metal chain. The links are called *monomers* (the Greek word for "single part"). Each link is made up of a particular arrangement of atoms. *Polymerization* is a chemical reaction in which the molecules of a monomer are linked together to form larger molecules. Polymers can also be produced by the polymerization of two or more polymers, which would then be called *copolymers*. *Copolymerization* makes it possible to create plastics for specific requirements by changing the arrangements of units (monomers) in the copolymer.

The following molecular diagram shows the ethylene monomer C_2H_4, which has been changed through polymerization into a long chain of molecules to form polyethylene. The monomer is the repeating structural unit, indicated by a box in the diagram, which in this case contains two atoms of carbon and four atoms of hydrogen.

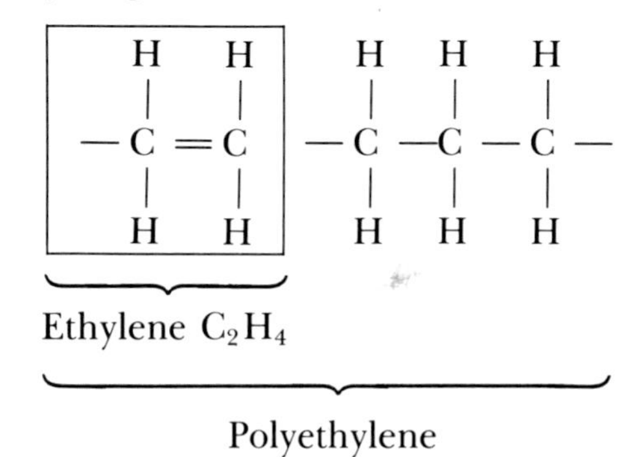

CLASSIFICATION OF POLYMERS

There are thousands of variations of commercial polymers (resins), making it impossible to list all of their properties. But plastics are grouped within families, and all the members of a given family have certain characteristics in common.

Polymers are classified into two groups: thermoplastic and thermosetting. *Thermoplastic polymers* are softened by heat and can be shaped and molded. Some thermoplastic materials burn when exposed to an open flame; others do not. Since there are no strong bonds between the individual molecules, they can be molded over and over again. This is a great advantage from the standpoint of recycling, and, in fact, most commercial plastics are thermoplastics.

Thermosetting polymers are created by crosslinking molecular structures, thereby creating strong chemical bonds between polymer chains. These materials are hard and rigid. They do not soften when heated and are not flammable.

Commercial polymers within each group, or family, and their characteristics follow.

Thermoplastic Polymers

Polyethylene is probably the most familiar of all plastics. It is available both as a film and as a flexible plastic and is used for many purposes, including containers, bottles, toys, housewares, plastics bags, tubings, and coatings.

Polyethylene is produced from ethylene gas, a by-product of petroleum. It can be injection molded, blow molded, extruded, vacuum formed, casted, and calendered; it can be either high or low density.

Polypropylene was developed in 1954 by Guilio Nata, a professor at the Polytechnic Institute of Milan. It is produced from propylene gas, another by-product of petroleum. Polypropylene is a strong, hard, white material that is resistant to cracking. It is used for luggage, cosmetics, containers, and automobile and aviation components. Because it can be sterilized it is also used in hospital equipment. Polypropylene is most suitable for blown bottles, housewares, electronic parts, and fibers for carpeting.

Polystyrene was first synthesized in 1866 but was not produced commercially until the 1940s. Synthetic styrene rubber BUNA-S was developed during World War II, and polystyrene film was introduced in 1948.

Polystyrene is inexpensive and therefore is used for disposable items such as cups, containers, trays, and utensils. Typical molded, polystyrene products include appliance parts and housings, furniture, optical components, toys, and hobby kits. In fact, polystyrene sheets are the favorite material of designers and modelmakers because they are easily cemented.

Another popular form of polystyrene is the expanded foam (Fig. 4-3). It is used in packaging fragile objects, in insulated products, and as insulation material for the building and construction industries. It can be applied in sheets in liquid form.

Styrene acrylonitrile (SAN) is produced by copolymerizing acrylonitrile and styrene to form the SAN polymer. Styrene acrylonitrile is used for food packaging,

Figure 4-3. Expanded polystyrene products. *(Photo from F.I.T. Packaging Design Workshop.)*

Figure 4-4. Telephone made from acrylonitrile-butadiene-styrene (ABS).

glasslike containers and bottles, lenses, electronic components, and piano keys.

Acrylonitrile-butadiene-styrene (ABS) is a *terpolymer* (a triple polymer) that combines flexibility with toughness. Products typically made of ABS are boat hulls, automobile trims, telephones, power tool housings, football helmets, radio and television cases, extruded pipes, and pipe fittings (Fig. 4-4). Acrylonitrile-butadiene-styrene is suited to most molding and calendering processes.

Polyvinyl chloride (PVC) is the most important polymer in the vinyl group. Polyvinyl chloride is produced from acetylene and hydrogen chloride. It has excellent resistance to water and chemicals and is self-extinguishing. Among its applications are wire coatings and glasslike blown bottles. Calendered PVC sheets are used as simulated leather (i.e., vinyl), shoes, handbags, rainwear, automobile interiors, shower curtains, coats, and upholstery materials. Polyvinyl chloride pipes and tubings are widely used in the building industries. Products made of PVC are processed by means of extrusion; blow, injection, and rotational molding; and calendering.

Polyvinylidene chloride is prepared from ethylene chloride and is available in both rigid and flexible forms. It is widely used in upholstery, pipes and tubing, and outdoor furniture. As a film, it is an excellent material for food wrapping and is familiar to consumers as Saran Wrap™. It can be molded, extruded, and calendered.

Polyvinyl chloride plastisol, a liquid form of PVC, is basically a coating material. When heated, the resin swells and forms a solid coating material. Polyvinyl chloride plastisol is used in molding soft, flexible products and as a coating for metal parts.

Polycarbonate resins are among the toughest of all plastics. They can be used in place of many metal parts and are found in electrical panels, insulators, street lighting globes, football and safety helmets, machine parts, shoe heels, electric appliances, and power tool housings (Fig. 4-5).

Cellulose acetate is produced by extrusion, injection molding, or compression molding. In sheet form it is used in vacuum forming. Other applications include toys, combs, lampshades, and housewares. The extruded film is used in X-ray, photographic, and tape-recording equipment.

Figure 4-5. Safety helmet made from polycarbonate resin.

Cellulose acetate can also be made into fibers for weaving into a fabric known as acetate rayon.

Cellulose propionate is usually processed by means of extrusion or injection molding. This economical material is used in items such as toothbrush handles, pen and pencil barrels, and tool handles.

Cellulose acetate butyrate (CAB) is the "outdoor" plastic; low temperatures do not diminish its strength. Extrusion and injection molding techniques are used to form CAB into sheets, tubing, outdoor signs, mailboxes, and building materials.

Ethyl cellulose has several outstanding properties, including high impact strength at subzero temperatures, rigidity, and suitability for molding. It is used in bowling pins, safety helmets, gears, and outdoor objects.

Cellulose nitrate is rarely used today because it is flammable. It is found primarily in explosives and pigments.

Cellophane is a pure form of cellulose. It is produced by treating cellulose from cotton or wood pulp so that it dissolves. The solution is spread onto flat sheets, which pass through a bath in which the pure cellulose is reformed or regenerated. The film then passes through a plasticizing solution, making it less brittle, and is coated with lacquer, which makes it moisture proof. Cellophane is used mostly for wrapping food products and in die-cut window boxes (to face direct holes).

Nylon™ is the trade name for a group of polyamide resins introduced by the du Pont Company in 1938. A du Pont chemist, Wallace H. Carothers, had been experimenting with condensation polymerization, and his work led to the development of polyamides. Nylon was originally used as a fiber for fabrics and did not come into widespread use as a molding material until the 1950s. Nylon can be injection molded, blow molded, and extruded. Its major advantage is strength. It is used in gears, bearings, ship propellers, hinges, fishing lines, and textiles.

Acrylic (methyl methacrylate) polymers are produced from petroleum-based ethylene and propylene. The outstanding properties of acrylic resins are exceptional clarity and the ability to transmit light. Acrylic rods and fibers are widely used in diagnostic medicine to conduct light from a single source (fiber optics). Acrylic plastics appear under numerous trade names, including Lucite™, Plexiglas™, Acrylan™, Orlon™, and Dynel™. The last three are fibers suitable for all kinds of clothing. Water-soluble acrylic paints are used by artists and decorators.

Polyphenylene oxide (PPO) offers outstanding electrical properties over a wide temperature range. Many uses have been found for PPO in electronics. Small electrical appliances, heaters, switches, battery cases, printed circuits, and electrical housings are made from PPO. It can be sterilized and is used in making medical and surgical instruments.

Acetal resins are prepared from formaldehyde. They are highly resistant to most chemicals and can be easily processed using all molding and extrusion methods. They are used by the automobile industry in carburetors and fuel pumps and are suitable for hardware items and aerosol bottles.

Polysulfone resins are produced for high-temperature applications. Owing to the high heat resistance of this plastic, it can replace many thermosetting materials. Polysulfone can be adapted to many molding processes. Product applications include circuit breakers, microwave oven parts, automobile distributor caps, and medical equipment that can be sterilized.

Polyurethane can be prepared as either a rigid or as a flexible, pliable material. Most polyurethane is produced in the form of foamed plastic. Rigid polyurethane resins called elastomers can be stretched to more than twice their original size. They are used in tires for heavy equipment and for automobile bumpers (Fig. 4-6).

Figure 4-6. Rigid polyurethane tire.

Polyurethane foams are available as liquid sprays that can be used for coating and packaging fragile objects. They also come in the form of stock slabs, which are used for furniture cushioning, insulation materials, sponges, life jackets, filling in aircraft wings, and insulation for fuel tanks on the space shuttle.

Polytetrafluoroethylene (Teflon™) and *tetrafluoroethylene* resins (TFE) are the most inert of all plastics. They are extremely resistant to solvents and corrosive chemicals, but because of their high melting point they are hard to work with. They can be molded as a powder, then fused. Many consumer and industrial products make use of these resins. Products include nonstick cookware, irons, and frying pans; piston rings; bearings; gaskets; and chemical-resistant tools.

Ionomer resins are produced from ethylene gas. They are tough and highly resistant to chemicals. Ionomers can be injection molded, blow molded, thermoformed, or extruded. They are used in molding bottles, sports equipment (e.g., golf balls), and containers for food packaging. Papers and fabrics can be coated with ionomer film.

Polyimide resins can be formulated in thermoplastic as well as thermosetting methods. Polyimide is one of the most heat-resistant plastics. It is widely used in the electronics and aerospace industries to insulate wires and motor windings in aircraft and missiles.

Polyallomer resins are produced from ethylene and propylene. Since they are light in weight and flexible, they are used in hinges for luggage, cosmetic containers, flex-top packages, closures, and applicators. Polyallomers are processed by extrusion, injection molding, or thermoforming.

Thermosetting Polymers

Thermosetting polymers differ from thermoplastic polymers in many ways. As noted previously, thermosetting polymers are cross-linked networks of long chain molecules; hence, they are rigid and infusible. Once they have been molded into a shape, they cannot be reheated and reshaped. Compression molding is the chief method used to make thermosetting plastics.

Phenol formaldehyde resins, commonly known as *phenolics*, are rigid, heat-resistant materials. These inexpensive plastics are produced by means of compression and transfer molding. They are used in a liquid form for laminating plywood, fabrics, and board. Phenolics are also used in brake linings, distributor caps, automobile body parts, and housewares.

Polyester resins are available in liquids, solids, pastes, and fibers. When they are combined with glass fibers, the result is *fiberglass*, which is used in boat hulls, aircraft and automobile bodies, gears, luggage, and tools. The resins are compression molded. Soft-drink bottles made from polyester-oriented polyethylene terephthalate (PET) are blow molded after being preshaped by injection molding. Polyesters in liquid or paste forms are mixed with catalysts at the time of use. Catalysts in this case are hardeners of adhesives, filters, and other plastic solvents, glues, and powders; an example is epoxy glue.

There are also thermoplastic polyesters that can be injection molded and extruded. They include two well-known synthetic fibers: Dacron™ and Fortrel™.

Amino resins include melamine formaldehyde and urea formaldehyde. These clear thermosetting resins are fireproof and resistant to detergents and oils. Molded products made from urea include buttons, closures, TV and radio cabinets, and switch plates. The best-known melamine product is china-like dinnerware. Both urea and melamine are used as adhesives in wooden products.

Alkyd resins are produced as molding compounds and liquids. About 90 percent of alkyds are used in liquid form as coatings. Alkyd resins in liquid solutions are used in odorless paints, enamels, and lacquers. As molding compounds alkyds are compression molded to form electronic components.

Epoxy resins are used in coatings to resist corrosion of pipes, tanks, containers, wall finishes, steel, and masonry. Epoxy adhesives are very strong and can be adapted to metals, glass, ceramics, and some plastics. Molding compounds of epoxy are used with catalysts for compression and transfer molding of electronic components and boat and aircraft bodies.

Allyl resins include diallyl phthalate (DAP) and diallyl isophthalate. These materials are used in rocket and missile components. They are molded by means of transfer and compression processes.

Silicone resins are chemically inert, odorless, and nontoxic; they are resistant to most chemicals, oils, and radiation. As molding compounds, they are suitable for transfer and compression moldings. As liquids, they are used for coatings. Molded, flexible silicones are used in airplane parts and components, artificial heart valves, gaskets, and small electronic parts. Silicone compounds are used in the aerospace and appliance industries to seal electronic parts.

TECHNIQUES USED TO MOLD AND SHAPE PLASTICS

As clear from the discussion so far, several different processes are used to convert polymers from their pellet, powder, or liquid form into plastic products. In this section we will describe several of the most commonly used processes.

Molding Processes

Injection molding is a widely used process to form thermoplastics. It is illustrated in Figures 4-7 and 4-8. First, a hopper is loaded with resin; then heat is applied until the plastic becomes soft enough to flow. The liquid plastic is forced through a nozzle into a mold. After the plastic has cooled, the mold separates and ejects the product. Production by injection molding is very fast, taking only 10-30 seconds per item.

Specialized equipment is used to mold thermosetting plastics. Injection molds

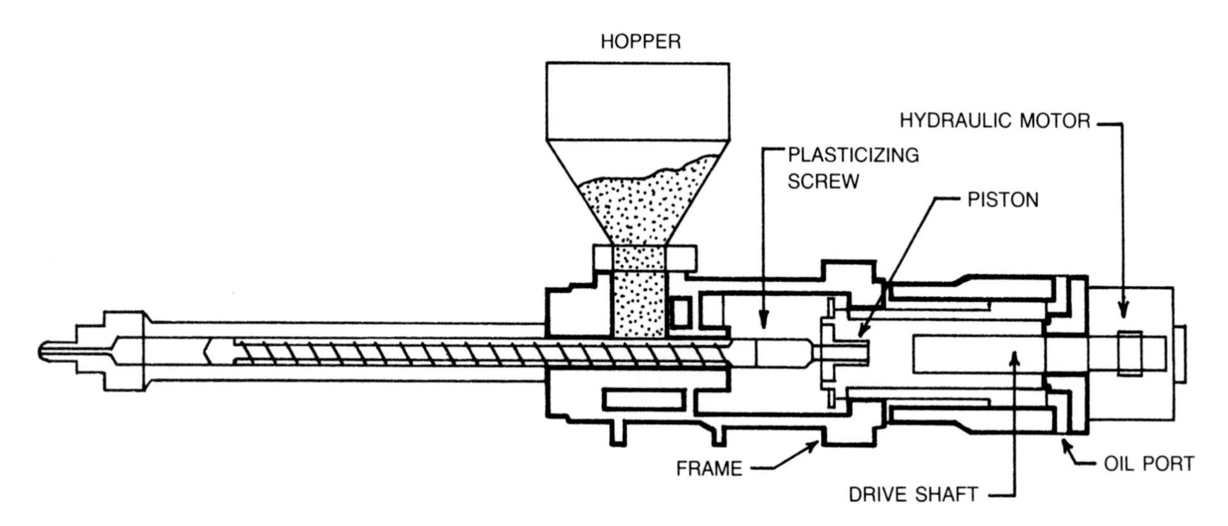

Figure 4-7. A reciprocating screw injection molding machine. *(Diagram by Elizabeth Downey.)*

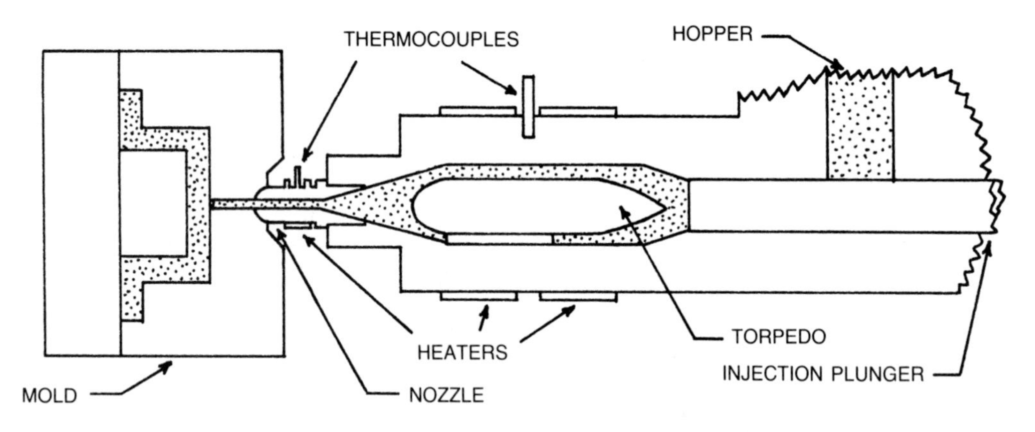

Figure 4-8. A plunger type injection molding machine (detail). *(Diagram by Elizabeth Downey.)*

Figure 4-9. Injection molding machine. *(Photo courtesy of Reed-Prentice Division Package Machinery Company.)*

usually consist of two halves, one stationary and the other movable. Figures 4-9 and 4-10 illustrate the process of injection molding thermosetting plastics. There are also multiple-piece molds for complex parts. Since molds are expensive, this process is most appropriate for large production runs.

Blow molding is used to make hollow shapes. In this process a thin-wall plastic tube called a *parison* is placed in the mold, which closes on the softened polymer and pinches off the bottom. A jet of air blows into the other end and expands

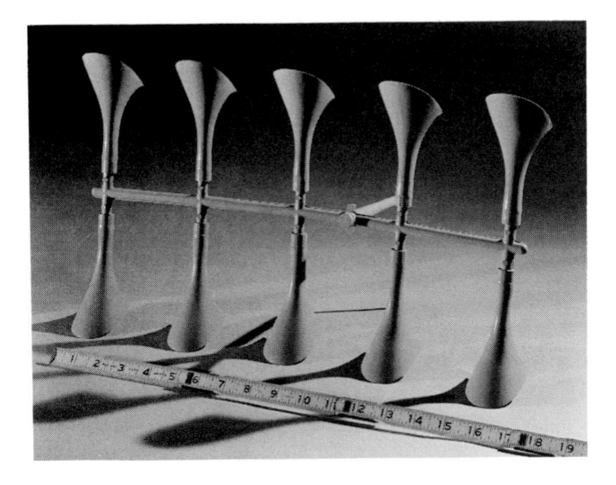

Figure 4-10. Injection molded plastic heels. *(Photo courtesy of Reed-Prentice Division Package Machinery Company.)*

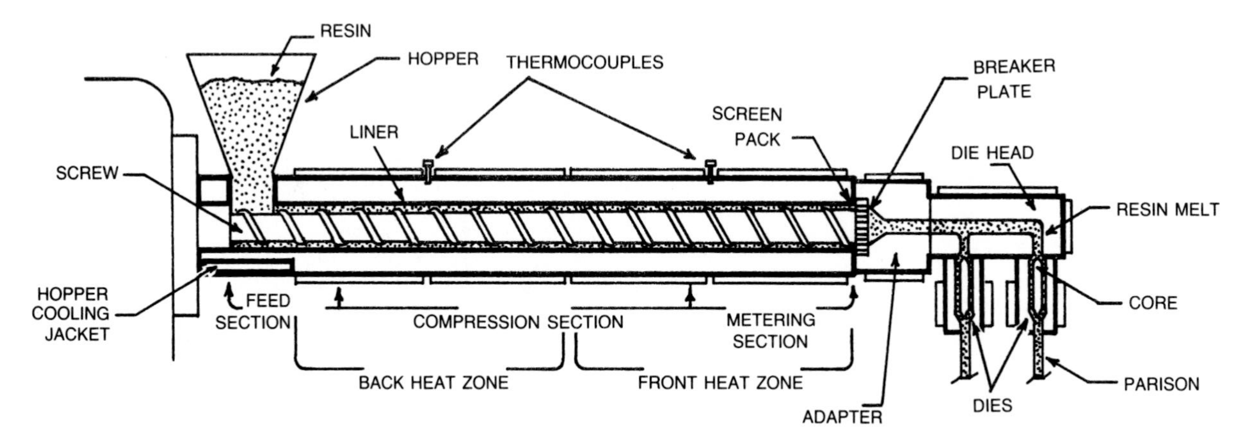

Figure 4-11. Blow molding. Cross section of extrusion and die units of a twin-head blow molder. *(Diagram by Elizabeth Downey.)*

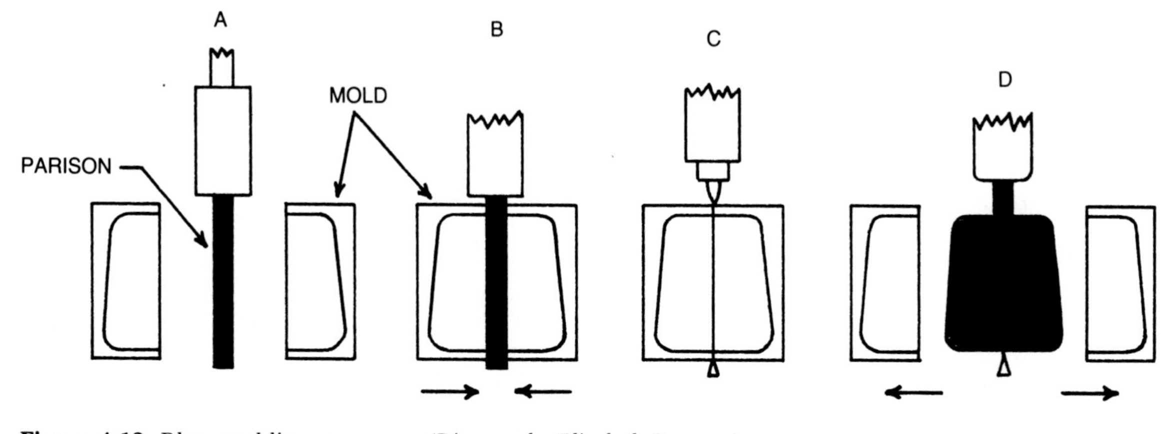

Figure 4-12. Blow molding sequence. *(Diagram by Elizabeth Downey.)*

the plastic to fill the cavity. The mold is then opened and the product ejected (Figs. 4-11 through 4-14).

In *compression molding*, thermosetting resin is placed into the heated mold (Fig. 4-15). Pressure is applied and the molten resin fills the mold cavity. The mold can have one or more cavities.

Transfer molding involves the same principle as compression molding except that the resin is not placed in the mold cavity. Instead, it is placed in a separate cavity known as a *transfer pot*. There it is heated under the pressure of a plunger, which forces the softened resin into the mold cavities. This molding process is most suitable for small, intricate parts.

Figure 4-13. Blow molding machine. *(Photo courtesy of Hoover Universal, Inc.)*

Figure 4-14. Blow molded polyethylene, polystyrene, and PVC bottles. *(Photo courtesy of W. Braun Company, Inc.)*

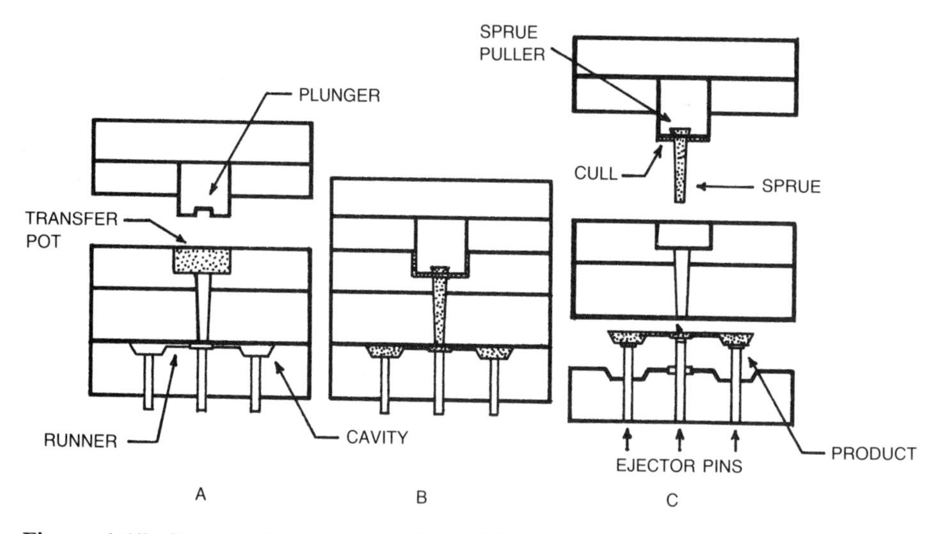

Figure 4-15. Compression and transfer molding. Diagram of the transfer molding cycle. *(Diagram by Elizabeth Downey.)*

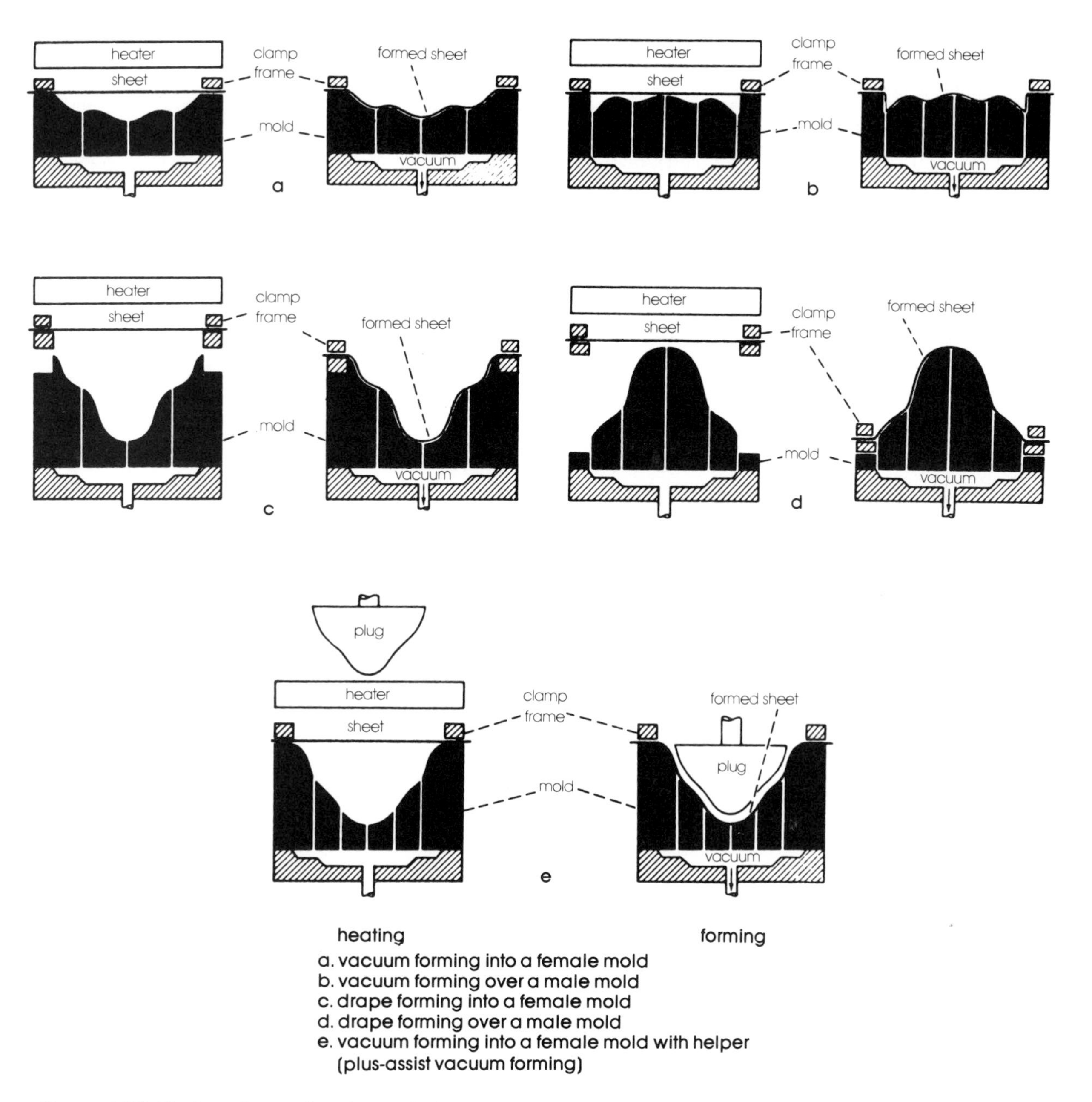

Figure 4-16. Various thermoforming techniques.

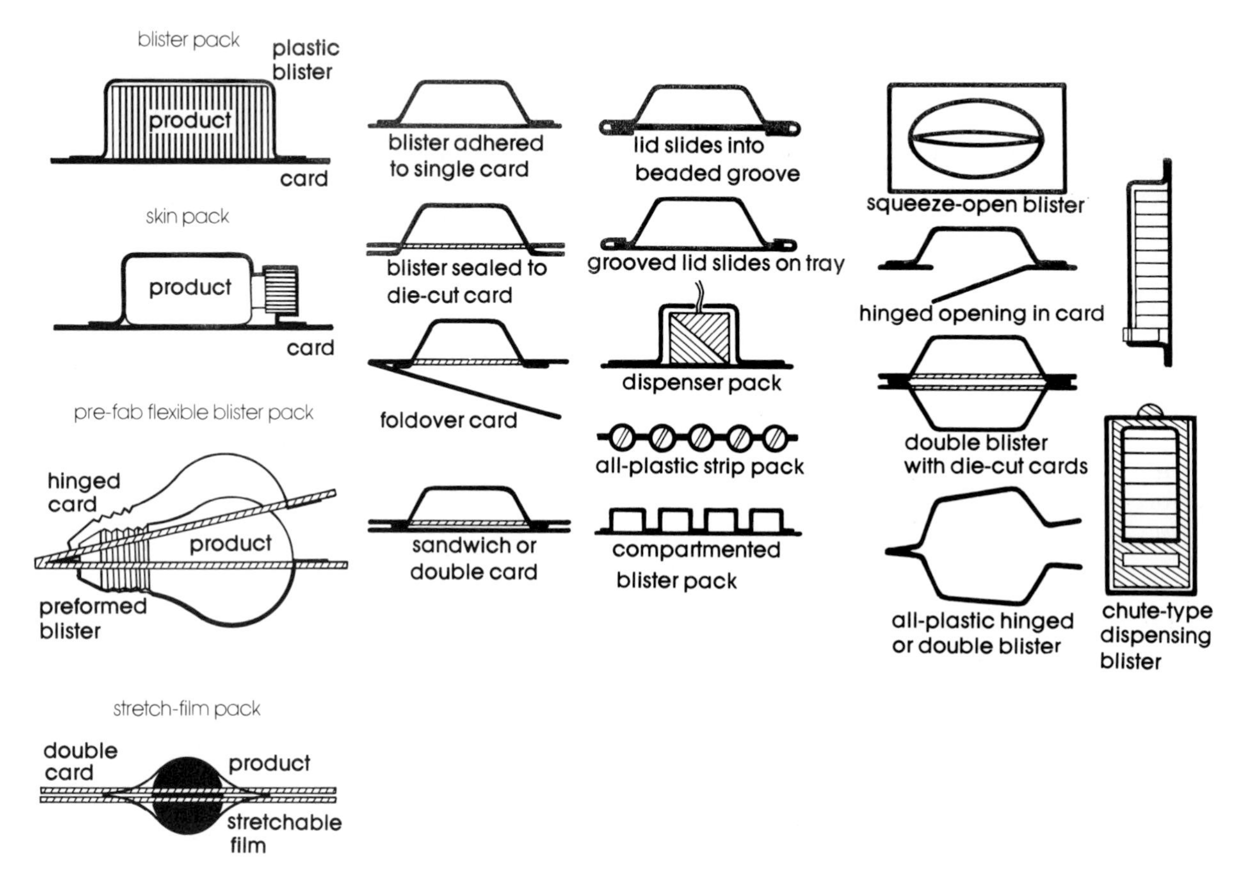

Figure 4-17. Basic thermoforming and card pack constructions.

Thermoforming or vacuum forming is the primary technique used for shaping plastic sheets (Figs. 4-16 and 4-17). The sheet is held in a frame and heated until it is soft and pliable. The mold is placed under the heated sheet, and a vacuum is applied through small holes in or around the mold. When the plastic cools, it hardens and retains the shape of the mold. Several mold systems are available, including those shown in Figures 4-18 and 4-19.

Figure 4-18. Small thermoforming machine. *(Photo from F.I.T. Packaging Design Workshop.)*

Figure 4-19. Large production equipment for thermoforming. *(Photo courtesy of Packaging Systems Corporation.)*

Figure 4-20. Thermoformed (blister) packaging. *Left to right:* Julie Herbig, Dawn Bellomo, and Erika Stalzer, designers.

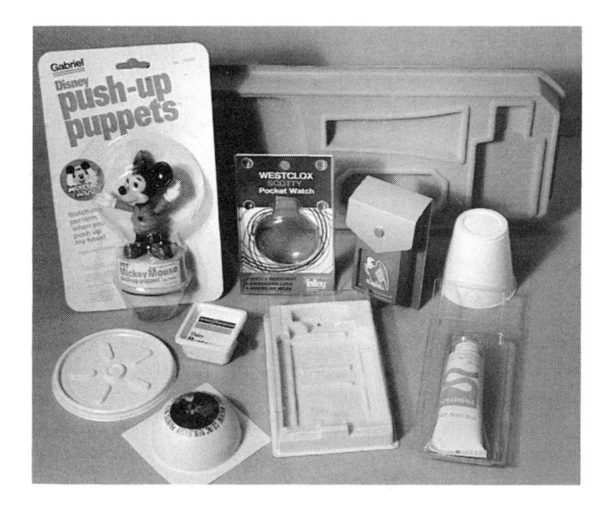

Figure 4-21. Thermoformed objects. *(Photo from F.I.T. Packaging Design Workshop.)*

Thermoformed packages are in widespread use and take many forms. Among them are the following:

Blister pack. This is the most widely used of all card packs and is intended to provide visibility (Figs. 4-20 and 4-21).

Prefabricated blister pack. A preformed blister is attached to a die-cut, hinged window card. The product is displayed inside the bubble and is visible on both sides of the card.

Skin pack. The product is laminated to a card with a thin, tough film.

Stretch film pack. A film is bonded to a die-cut double window card so that the product is encased in a see-through display package.

Thermoforming offers some exciting options. In the hands of a creative and informed designer, it can be a highly effective self-service packaging form for a wide variety of products. But as can be seen from the diagrams, many factors must be taken into account when planning and developing a thermoformed package.

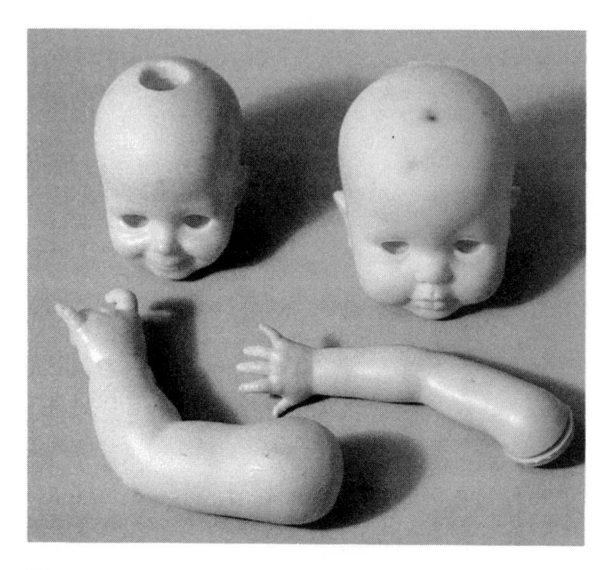

Figure 4-22. Plastisol slush-molded doll parts. *(Photo courtesy of Ideal Toy Coporation.)*

Another type of molding process is *plastisol molding. Plastisols* are mainly coating resins consisting of PVC resins and plasticizers. They are used to coat or dip nonplastic objects. Among the products produced by dip-molding are toys, spark plug covers, wire racks, tool handles, and housewares. Cold-dip plastisols are available for coating without heat and require only air drying.

In *slush molding*, plastisols are used to produce hollow doll parts and flexible toys (Fig. 4-22). In this process a preheated hollow mold is filled with liquid plastisol. As soon as the material begins to gel, it is dumped to drain off the excess liquid plastisol. After cooling, the mold is opened and the product removed.

In *rotational molding*, plastic resins or liquids are used to form large objects in a

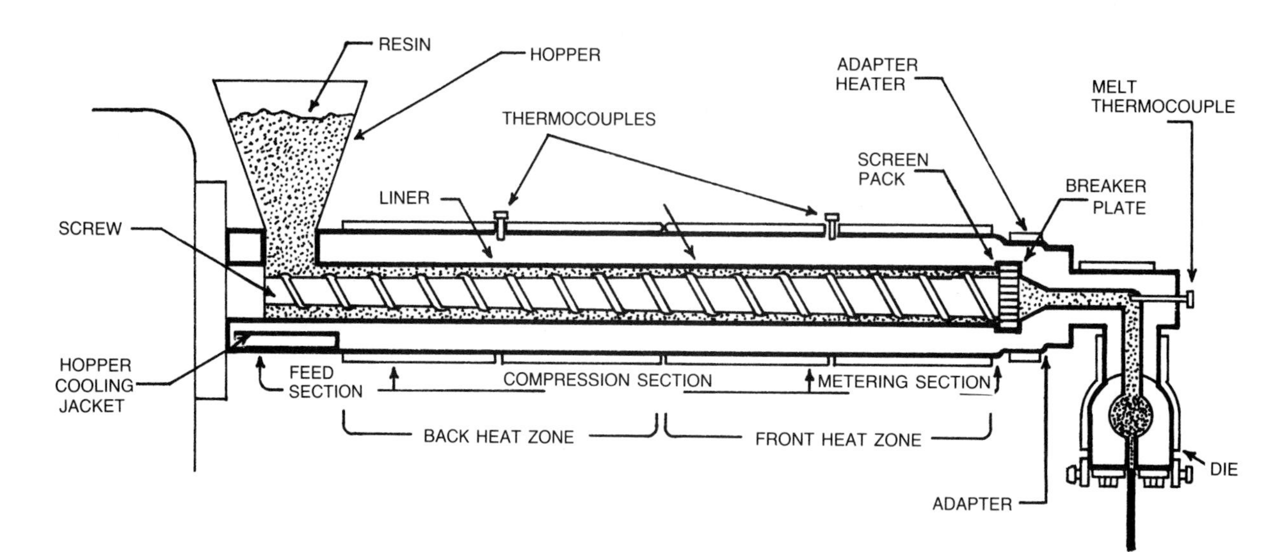

Figure 4-23. Extrusion. Cross section of an extruder. *(Diagram by Elizabeth Downey.)*

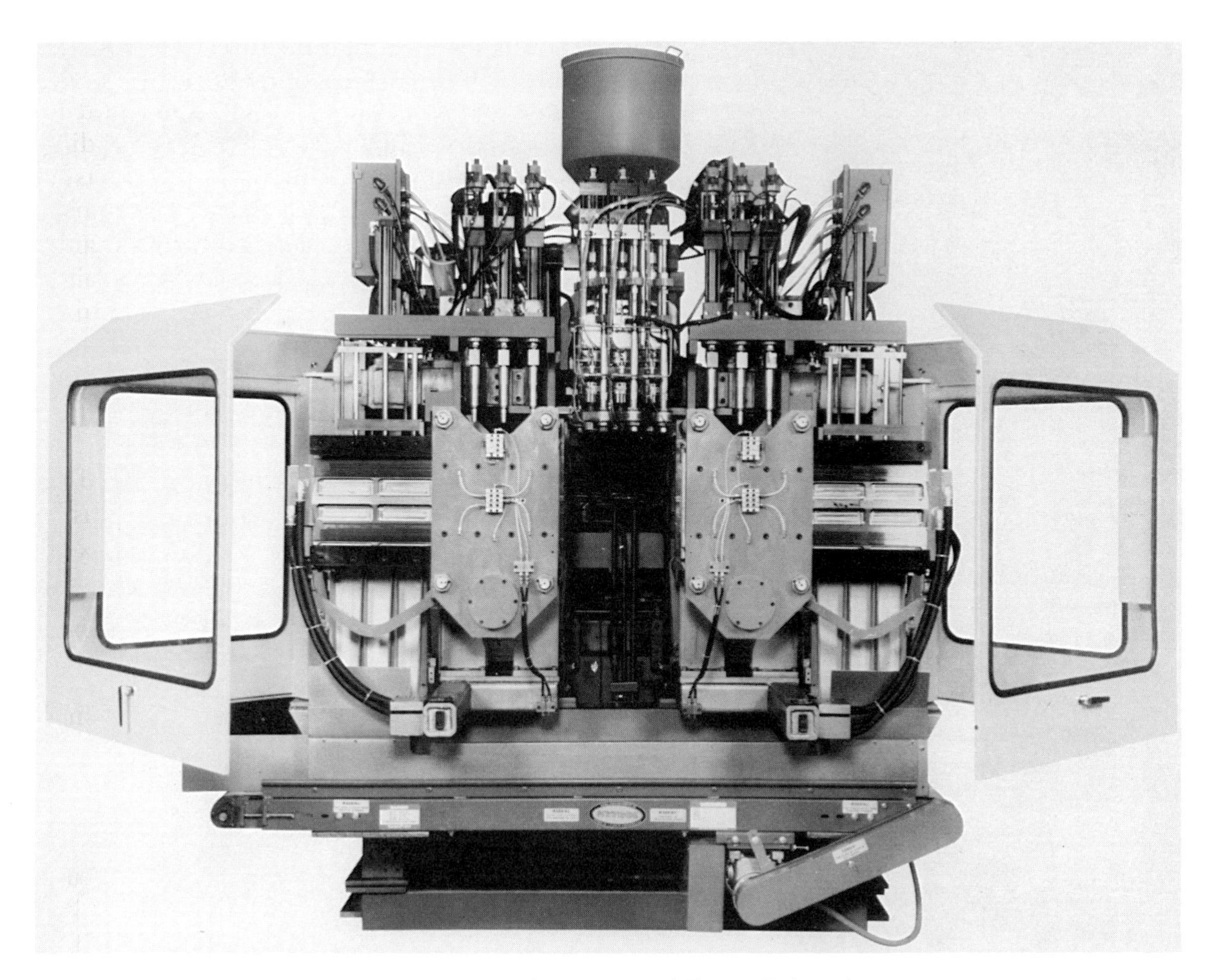

Figure 4-24. Continuous extrusion machine. *(Photo courtesy of Hoover Universal, Inc.)*

hollow mold that is rotated. After all the resin has fused into the mold's walls, it is cooled and the product is removed. Large (500-gallon) containers can be molded using this process. It is a slow process, but it is less costly than injection or blow molding.

Extrusion

In *extrusion*, thermoplastic resin is fed into a heated tube. The soft plastic is then forced through a die by a rotating screw or plunger to form a continuous shape (Figs. 4-23, 4-24, and 4-25). Extru-

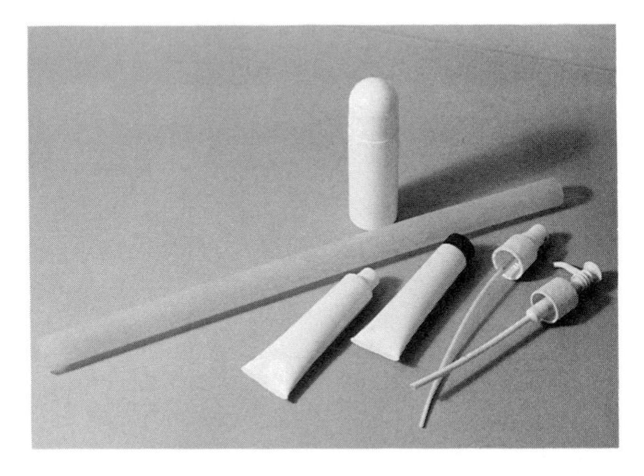

Figure 4-25. Extruded plastic products. *(Photo courtesy of Hoover Universal, Inc.)*

sion is used to manufacture rods, pipes, sheets, film, and coatings for cables and wires. This process is used only with thermoplastic resins.

In *coextrusion*, two or more films are simultaneously extruded to produce a multilayered film. Coextruded films are used in packaging for frozen foods, cereals, and meat.

Calendering

In *calendering*, softened thermoplastic material is pressed between two or more rolls (cylinders) to form a continuous sheet. Figure 4-26 shows various stages of the calendering process. Calenders are also used to coat papers and fabrics.

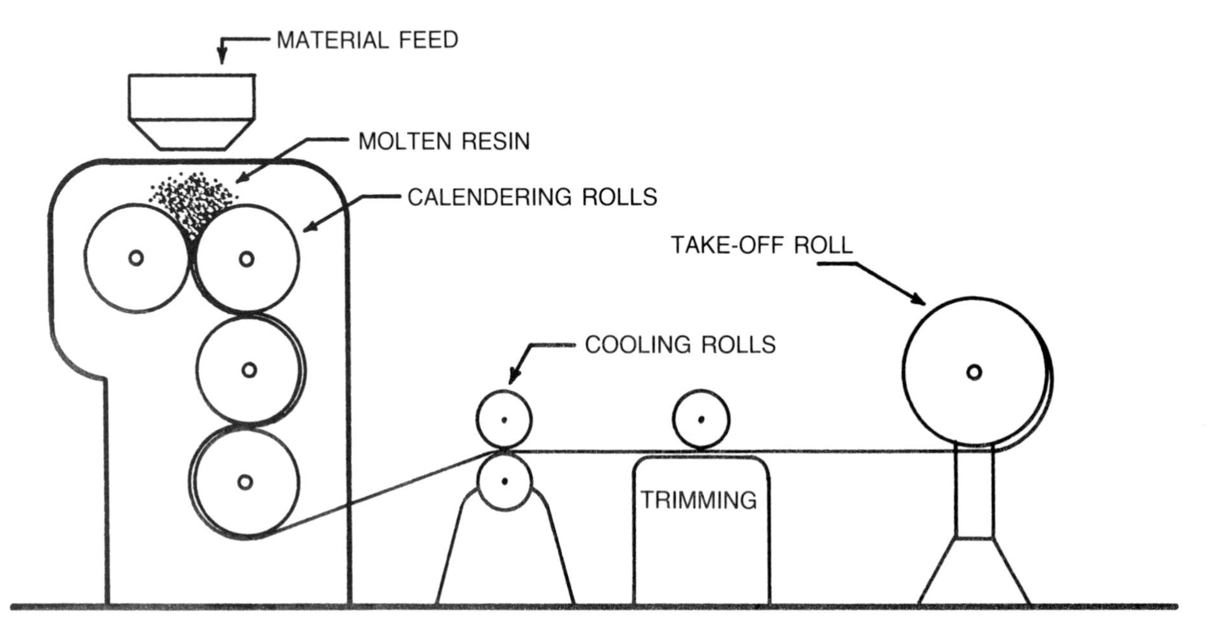

Figure 4-26. Calendering. The main stages in the calendering of plastic films. *(Diagram by Elizabeth Downey.)*

Laminating

Laminating is a process in which two or more layers of material are bonded together to form a single sheet. Plywood and paperboard are examples of products that are made by this process. Most lamination is done with a heated hydraulic press.

Casting

Casting with plastics is similar to casting with other materials, such as plaster, cement, and latex. The liquid plastic is poured into the mold. The plastic is a liquid resin to which a catalyst (hardener) has been added. The mold may be a simple plaster mold or a complicated precision mold with many parts. Resins used in casting include acrylics, polyester, silicones, epoxies, nylons, phenolics, and urethanes.

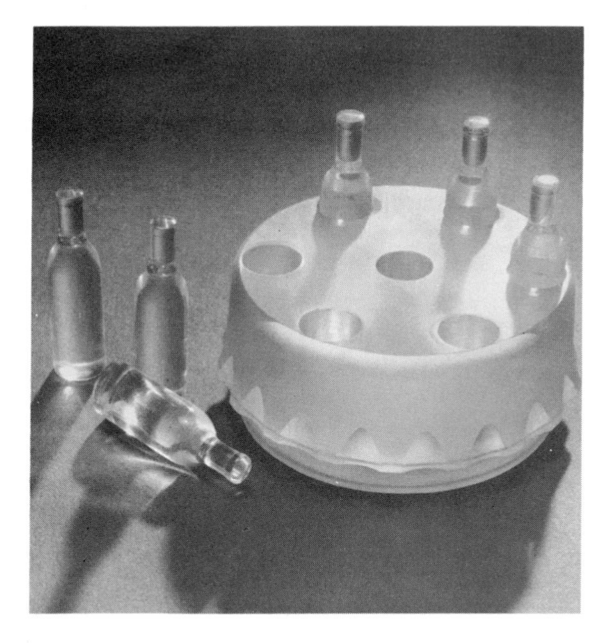

Figure 4-27. Fabricating with plastics. "Spin the Bottle" game product model made of polystyrene and lucite. Frank Csoka, designer. *(Photo courtesy of Fun Things, Inc.)*

Fabricating with Plastics

Fabrication consists of cutting, shaping, and bonding (cementing) materials to form objects (Fig. 4-27). Plastics may be joined either by adhesion or by cohesion. *Adhesion*, commonly known as pasting or gluing, sticks objects together. It makes use of layers of different materials, which stick to each surface of the joint to be fitted together. Contact cement resins are an example. *Cohesion* requires that the surfaces to be joined be softened, such as in welding, by means of a solvent. When the liquid evaporates, the joint becomes a single piece of plastic. Some plastics (e.g., polyethylene) are not easily cemented by cohesion. Others, such as polystyrene and acrylics, are easily cemented in this way. Thermosetting plastics are hard to dissolve and do not cement well by cohesion.

Heatsealing is a process in which layers of plastic films are melted together. It is used to form pouches and envelopes and to seal packages and seams. Figure 4-28 shows a hand, heat-sealing electric iron.

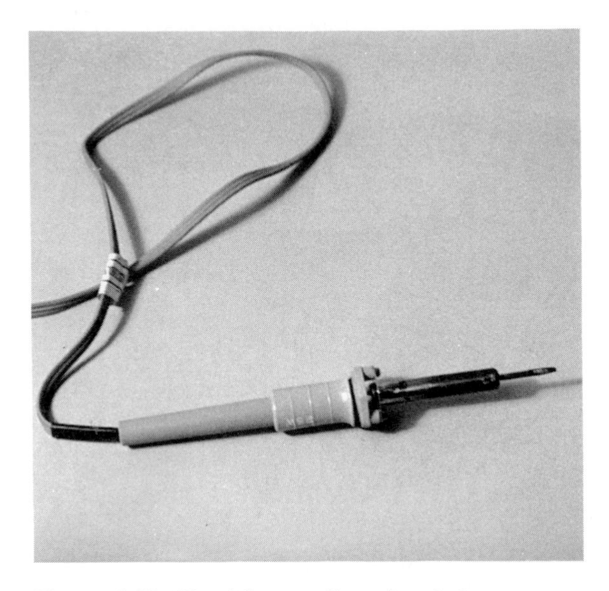

Figure 4-28. Hand heat-sealing electric iron.

Cutting Plastics

Plastic materials can be cut with a saw (preferably a mechanical saw), shears, or a sharp mat knife, depending on the thickness and type of plastic. Using these tools, they can be cut like any other material to fabricate objects such as boxes, furniture, frames, novelties, and so on. Plastic can also be drilled, sanded, and turned on a standard metal lathe.

DESIGNING WITH PLASTICS

The rapid development of plastics had a dramatic effect on packaging and design. Never before had product and package designers been offered so many opportunities to exercise their creativity. Today many consumer and industrial products are made of plastics; they can be produced at a lower cost than products made from natural materials.

In a free-enterprise, consumer-oriented society, increasing quantities of goods must be produced each year to satisfy the demand for an ever-higher standard of living. At the same time, consumers continually seek change and novelty, a tendency that is exploited to the fullest by marketers. New styling, new features, even new colors will make last year's model obsolete.

This "planned obsolescence" is often criticized, but it has numerous benefits. It increases the availability of most consumer products. It stimulates competition in the marketplace, often improving the quality of products and creating job opportunities for millions of people. Many of these benefits are related to the availability of plastics for use in product and packaging design.

A well-planned design program with plastics requires the teamwork of many experts, including product, packaging, and graphics designers; engineers; computer technologists; and model makers. The product designer, often called the industrial designer, needs a background in basic engineering. To use plastics effectively, he or she must know which materials are appropriate for particular functions. In addition, the industrial designer must be able to design a product that performs well, is attractive to the consumer, and is economical for the manufacturer. Creativity, precision, and skill in drafting, rendering, and model making are required.

Once a design comp has been approved, the manufacturer takes over. Blueprints, mechanicals, and finished art are prepared for the purpose of making the mold (or die) for the product or package. Mold making is a costly, highly specialized, and very time-consuming job. It may take several weeks, even months, to prepare the mold that will produce a product from a carefully selected plastic resin.

Finishing and Decorating

A variety of processes are used to finish and decorate plastic products and packages. In general, the finishes include paint, print, coat, and texture. The specific processes follow.

Vacuum metalizing (Fig. 4-29) consists of coating plastic with a thin layer of metal to give it a metallic appearance. This method is used to create metallic surfaces for toys and automobile components.

Electroplating is another method for creating metal-like plastic products and components. Plumbing supplies, and automo-

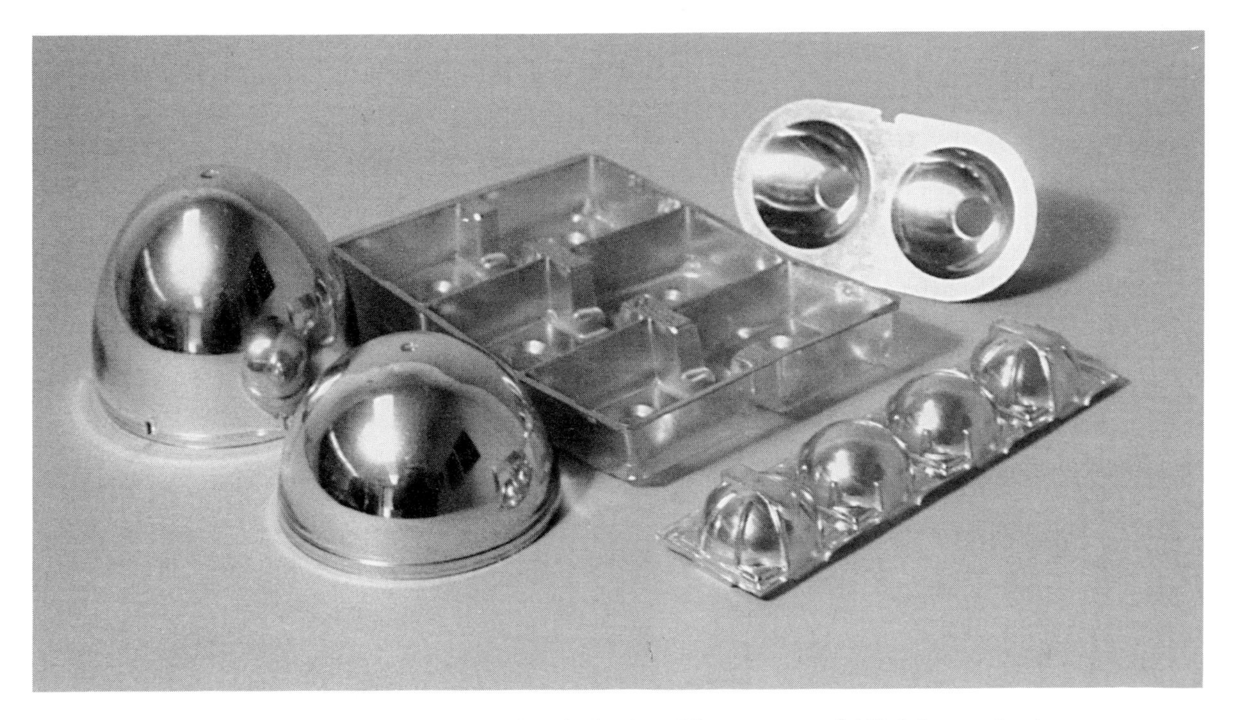

Figure 4-29. Vacuum metalized and electroplated plastics. *(Photo courtesy of Allied Corporation.)*

Figure 4-30. Plastic toy model. *(Photo courtesy of Ideal Toy Corporation.)*

bile grilles and taillights are made from chrome-plated plastics.

Hot stamping is a popular way of decorating plastics or cardboard for packages and products. The process consists of transferring the design from a thin foil or film onto the product. The application of heat and pressure to the back of the foil adheres the design to the product or package (Fig. 4-30). Several colors can be hot stamped in a single operation. Plastic bottles and cosmetic containers are hot stamped in metallic or nonmetallic colors.

Silkscreen printing is another method for decorating plastic products and packages. Plastics can also be *engraved* on a special engraving machine to produce signs and nameplates.

Some plastic products, especially dinnerware (melamine), are decorated using the *in-mold process* (Figs. 4-31A and B). A printed full-color overlay film is placed in the mold and fused to the molded piece during the molding process. This method is used in compression molding with thermosetting resins and can also be used in injection molding.

Figure 4-31A. In-mold process. Decorated Melamine dinnerware.

Figure 4-31B. In-mold process. Plastic objects. *(Photo courtesy of Autoroll-Dennison Corporation.)*

Plastic Bottles and Containers

It is important for designers to study the new plastic resins, especially their unique functions and product compatibilities. These can open up opportunities in bottle and container design (Figs. 4-32A and B). Understanding versatility and design possibilities of the new plastics is among the most challenging aspects of being a designer today.

Plastic offers certain advantages over glass, the main one being that plastic is lighter. In the past, problems of product compatibility prevented the use of plastic for bottles and containers. But compatible plastic materials have now been developed and are in widespread use. Liquid detergents, chemicals, and toiletries are being packaged in plastic containers (Figs. 4-33A and B). Plastic bottles are slowly infiltrating the food and beverage indus-

Figure 4-32A. Outstanding examples of contemporary design by Vignelli Designs for Castigliani and Heller Designs. *(Photo courtesy of Vignelli Designs.)*

Figure 4-32B. Vignelli Designs for Heller Designs. *(Photo courtesy of Vignelli Designs.)*

Figure 4-33A. Carboys. *(Photo courtesy of Bekum Corporation.)*

Figure 4-33B. A variety of stock plastic bottles. *(Photo courtesy of Hoover Universal, Inc.)*

try, and resins for bottle molding are continually becoming available.

Blow-Molding Equipment

The development of blow-molding equipment has made the plastic bottle a practical, inexpensive package. The procedure for designing plastic bottles and containers is similar to that used for glass bottles and jars. The designer working with plastics can, however, choose among a greater variety of materials and decoration processes. In addition, a variety of stock bottles and closures are available.

Stock Bottles and Closures

Plastic vials are small bottles, usually injection molded or fabricated from tubing. They are used to package small items, pills, capsules, and powders.

Pails are large plastic containers (e.g., garbage cans and trash bins). They are usually molded from high-density polyethylene.

Carboys are large, 3- to 15-gallon plastic bottles that are used for commercial liquid products and often encased in rigid outer containers (corrugated boxes).

Plastic cans are a recent development in soft-drink and concentrates packaging. They have also been introduced for petroleum products. Plastic cans are relatively inexpensive compared to tinplate cans.

Crates and shipping containers used to be made of wood. They were, however, heavy and deteriorated relatively quickly. Plastic shipping containers are now used

Figure 4-34. Clear plastic trays for produce.

to transport milk, soft drinks, and beer. Similarly, plastic drums are more versatile than the traditional wood or fiber barrel.

Other stock items include *trays* and *boxes*. A *tray* is a lidless container for carrying heavy objects. It can be a simple, divided platform like those in candy boxes or shaped to hold meat, produce, and other objects and sealed with plastic film (Fig. 4-34). Plastic trays and platforms can be thermoformed or injection molded from a variety of plastics, including polyethylene, polypropylene, polystyrene, PVC, acetate, and expanded polystyrene.

Early plastic boxes were fabricated containers, usually made of acetate or Plexiglas and joined with cements or solvents. Today most plastic boxes are injection molded and often combined with thermoformed platforms. Injection-molded rigid boxes are available in stock sizes in many styles and shapes. Like all plastic containers, they can be decorated by printing, hot stamping, screening, or labeling.

Model Making with Polystyrene

Designers often use polystyrene to make boxes, 3-D models, and other items. The necessary tools are a mat knife, some fine sandpaper, and a steel ruler. Also required is an adhesive, Rez-N-Bond™, which is available at plastic supply stores.

Polystyrene usually comes in 40″ × 72″ sheets and is surprisingly inexpensive. The thickness of the sheets varies from .020 to 250 points white opaque sheets. The sheets can be painted with enamels, oils, and acrylic colors or covered with fabrics or decorative papers.

The chief advantage of polystyrene is that it can be bonded in seconds with Rez-N-Bond. The bond is nontoxic. The adhesive is also suitable for adhering lucite, Plexiglas, and other acrylic plastics. Polystyrene can be scored with a mat knife and will snap away easily. It can also be bent with heat (e.g., the steam from a kettle) to create curved or rounded objects.

Packaging with Plastics

The packaging industry uses over 30 percent of all the plastics produced in the United States. And this percentage will increase in future years. Because of ecological pressures, the plastics industry has responded to environmental concerns by developing new methods for the disposal, recycling, and reuse of plastics. Manufacturers have also invested in the development of biodegradable plastics.

OTHER IMPORTANT USES OF PLASTICS

Plastics in Art

As early as the 1920s, artists produced sculptures, collages, and constructions using available plastics and plastic materials. The development of synthetic pigments

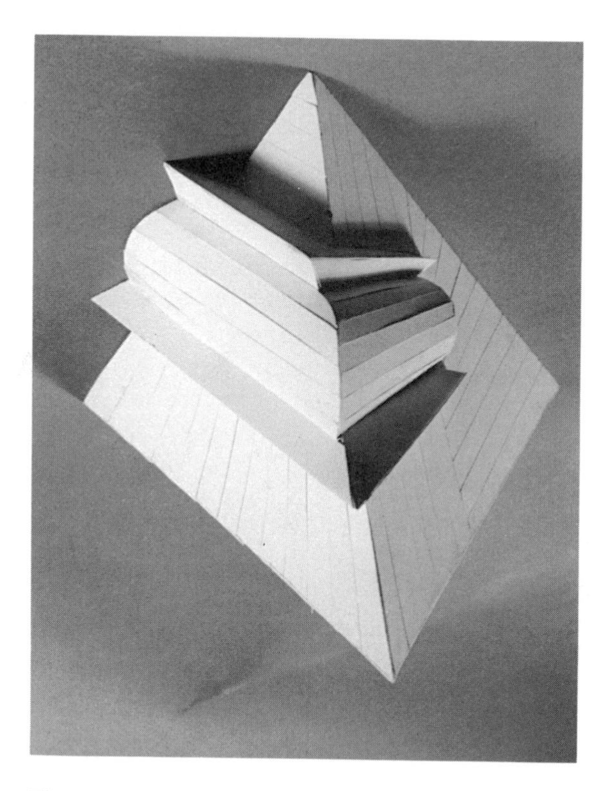

Figure 4-35. "Pyramid" polystyrene sculpture by Ivan Biro.

and paints in liquid, powder, and paste forms (acrylics) inspired painters and illustrators. Contemporary artists are using plastic media in both sculpture and painting (Fig. 4-35).

Plastics in Medicine

Medicine and dentistry have also benefited greatly from the availability of special plastic materials. Also, plastic valves, tubes, tubings, fiber optics, and pouches are frequently used in medicine and surgery. Prosthetic devices (replacements of limbs) are made of lightweight, flexible plastics. Disposable plastic instruments and accessories in sterile packaging are also in widespread use.

In recent years considerable progress has been made in the replacement of body parts and vital organs (implants). In orthopedic surgery, implants of synthetic bones and joints have become almost routine procedures. In ophthalmology, implants have saved the eyesight of many people. Recent progress in orthotopic cardiac replacement has led to the construction of the artificial heart. The ventricles are constructed of polyurethane, and the connections to major blood vessels are made of Dacron.

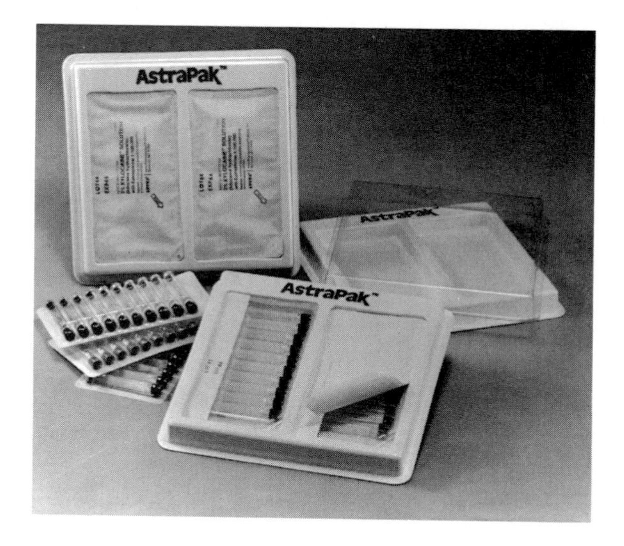

Figure 4-36. Anesthesia cartridges. *(Photo courtesy of Astra Pharmaceutical Products, Inc.)*

In dentistry, plastics play an important role in the restoration of teeth. They are also being used for adhesives and fillers.

Plastics are also important in pharmaceutical packaging. Disposable kits, medications, and other surgical-medical aids and products are supplied in sterile plastic packages (Fig. 4-36).

Plastics in Space Exploration

Plastics possess many qualities that make them ideal materials for use in outer space. Figure 4-37 shows an Extravehicular Mobility Unit used by NASA that is made of a variety of plastics and plastic sealers.

Figure 4-37. Extravehicular Mobility Unit. *(Photo courtesy of NASA.)*

PROJECTS

The first plastic package you will be asked to design will probably be a blister package or a skin pack. Begin by reviewing the descriptions and figures in this chapter. Then tackle the following projects.

1. To create a blister package, use the existing blister from a product card. Redesign the entire card, providing a new shape, new graphics, and a brand-new presentation.

2. To create a stretch film pack, devise a card and graphics for a product of your choice. Use a Saran-type film to enclose the product. You can stretch and seal the film with the heat from a hair dryer.

SUGGESTED READINGS

Fundamentals of Plastics. Cleveland, OH: Penton, 1969.

"Guide to Plastic Packaging," *Modern Plastics,* December 1987.

Hanlon, Joseph F. *Handbook of Package Engineering.* New York: McGraw-Hill, 1971.

The New England Journal of Medicine, February 2, 1984, pp. 312–314.

Packaging Systems Corporation. *Thermoforming.* Orangeburg, NY: Packaging Systems Corporation, 1978.

Baird, Ronald J., and David T. Baird. *Industrial Plastics.* South Holland, IL: Goodheart-Willox, 1982.

Du Bois, J.H. *Plastic History U.S.A.* Boston: Cahners, 1973.

Sacharow, Stanley, and Roger C. Griffin, Jr. *Basic Guide to Plastics in Packaging.* Boston: CBI, 1973.

5

FLEXIBLE PACKAGING

(Photo courtesy of Bloomingdale's.)

Types of Flexible Packaging / Shopping Bags

FLEXIBLE packaging has been in use since prehistoric times. Some of its oldest forms are baskets, bags, and wraps made from woven plants and animal skins (Fig. 5-1). Later forms include flexible paper packaging, the development of which came about during the American Civil War. Previously, cotton bags were used for storage and transportation of food. But after hostilities broke out and northern producers were cut off from sources of cotton, paper became the logical substitute (Fig. 5-2). Later, the mass-produced grocer's bag and paper sack were created by George West, a New York mill owner. These paper sacks were the ancestors of modern multiwall bags.

The introduction of cellophane and the development of plastic films, foils, and specialty papers, as well as methods of coating and laminating, led to the development of new types and uses of flexible packaging. Today the largest user of flexible packaging (80 percent) is the food industry (Fig. 5-3). Some of its packages—those designed for freezing, cooking, mixing, preserving, and dispensing foods and ingredients—are having a dramatic impact on the food industry, especially in the area of convenience (fast) foods.

Figure 5-1. Early flexible package (nineteenth century). Tobacco pouch. *(Photo courtesy of Landor Associates.)*

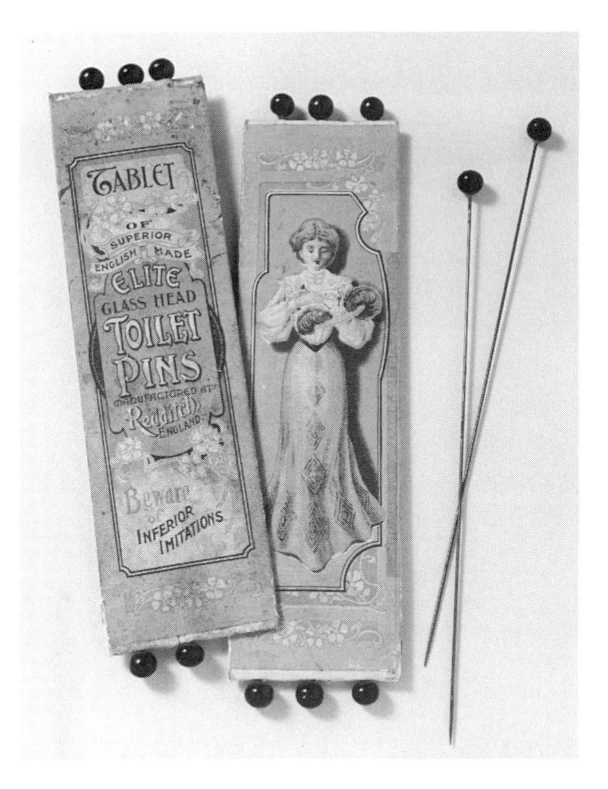

Figure 5-2. Early flexible package (nineteenth century). Knitting pins. *(Photo courtesy of Landor Associates.)*

TYPES OF FLEXIBLE PACKAGING

There are three basic types of flexible packaging: wraps and overwraps, preformed bags and envelopes, and form-fill-seal pouches. A fourth, recent, type is aseptic packaging.

Wraps and Overwraps

Both *wraps* and *overwraps* are sheets of flexible material that are usually fed from a roll stock. Wraps are formed around a product such as a candy bar or a loaf of bread; overwraps are formed around a basic package such as a carton. Variations on the wrap are bands and sleeves. Closures for wraps include adhesives, heat-seals, and peelable closures (wraps and overwraps can be seen in Fig. 5-3).

Various kinds of specialized high-speed wrapping machines have been developed, often with a high degree of automation.

Figure 5-3. Flexible packaging. *Left to right:* Richard Sarafian, Cindy Campo, and Lisa McGowan, designers.

paper styles

flat

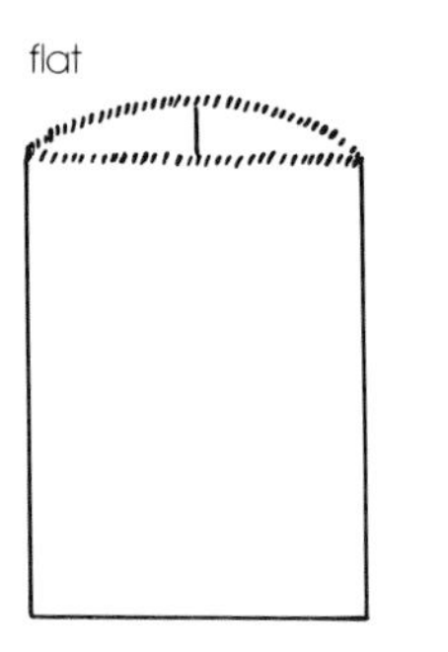

Has lengthwise back seam, no gussets. Bottom is generally folded over and pasted or heat sealed. Simplest bag style, most economical in use of materials.
Typical closures: Heat seal, folded and pasted, tape, tie, clip, staple.
Some uses: Potato chips, snack foods, ice-cream bars, soft goods, powered foods, coffee, frozen foods, hardware, machine parts.
Some features: Rack display at point of sale; billboard label area; fewer folds lessen water-vapor permeability.

automatic

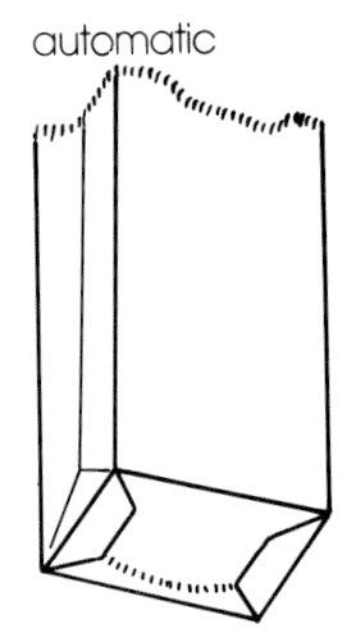

Self-opening style—SOS. Has side gussets, built-in flat bottom. It is rectangular in shape and self-standing.
Typical closures: Heat seal, folded and pasted, tape, tie, sewn.
Some uses: Coffee, cookies, dried milk, candies, insecticides, plant foods.
Some features: Has the best stacking ability, is easiest to handle. Stand-up display. Opens and fills easily.

square

Has side gussets (also called tucks or bellows folds). Bottom turned up and pasted or heat sealed.
Typical closures: Heat seal, folded and pasted, tape, tie, clip, staple.
Some uses: Fresh produce, snack foods, soft goods, candies, cookies, plant foods, insecticides, cereals.
Some features: Cubic capacity is greater than in non-gusseted bags of similar width. Side folds and bottom may be printed for greater display value no matter how bag is stacked.

satchel

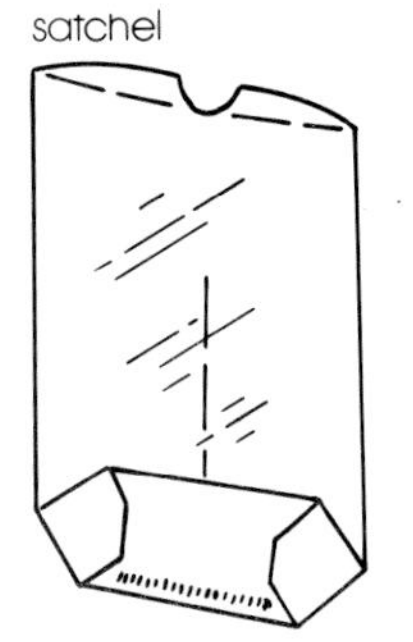

Has flat body, with no gussets. Bottom is formed in hexagonal shape, is flat and self-standing when filled.
Typical closures: Heat seal, folded and pasted, tape, tie, sewn, clip, staple.
Some uses: Sugar, flour, dried foods, cookies, baked goods, candies.
Some features: Combines economy of material with ease of handling and stacking. Stand-up display. Opens and fills easily.

Figure 5-4A. Types and uses of bags and pouches.

Preformed Bags and Pouches

A *preformed bag* is basically a tubular construction fabricated from paper, plastic, foil, fabric, or a combination of these materials. There are four standard styles of paper bags: flat, square, self-opening or automatic, and satchel. They are illustrated in Figures 5-4A, B, and C.

Invented around 1850, the *flat bag* is the oldest form of paper bag. Made by simply folding paper into a tube, it is used for hosiery and textiles. The *square bag,* in contrast, has pleats or gussets and is designed to hold bulky materials. Originally made of paper, it is now available in

film styles

bottom gusset

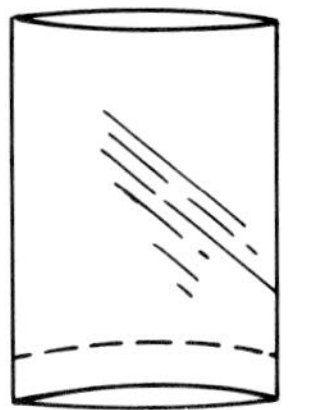

Has bottom gusset and side welds. Lip is optional, can be die cut with holes for wicketing. Made of thermoplastics, generally polyethylene.
Typical closures: Twist tie, plastic clip, tape, heat seal.
Some uses: Bread, paper goods, toys, multiple packages, household items, hardware.
Some features: Suitable for high-speed, automatic filling on newly developed bagging machines. Easy opening, reclosable, reusable. Offers visibility, strength, large-volume capacity.

flat-wicketed

Has side welds, no bottom seam or gusset. Made with or without lip. Lip can be die cut with holes for wicketing. Made generally of thermoplastics.
Typical closures: Twist tie, plastic clip, tape, heat seal.
Some uses: Produce, hosiery, candies, snack foods, hardware, toys, fresh meats.
Some features: Simple construction, visibility, easy opening, reclosable, strong, reusable.

specialty styles

contour

Two-compartment, foldover construction; suited to packaging multiple units, with advantage of enabling product inspection.

wallet

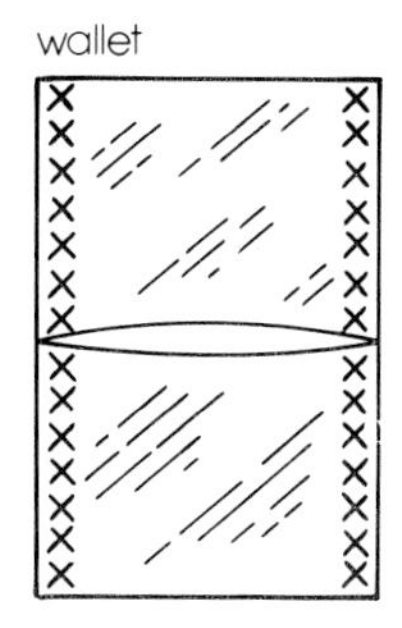

Fits rounded or irregular products: often made of shrink film or polyethylene. Used for phonograph records, poultry, fresh meats.

roll-fed

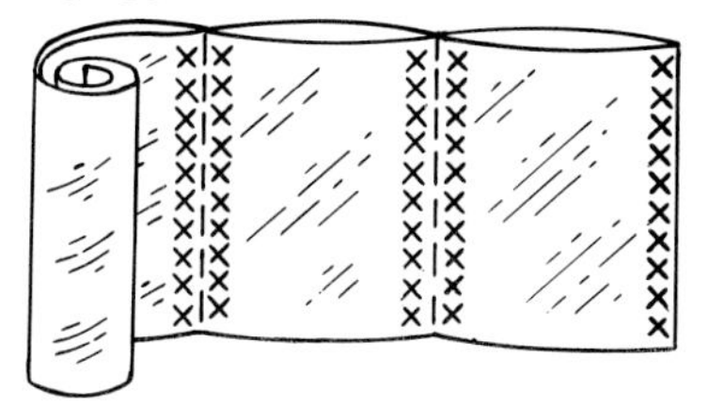

Thermoplastic film, often polyethylene, is perforated between bags for easy separation. Used in small-volume hand-filling operations.

Figure 5-4B. Types and uses of bags and pouches.

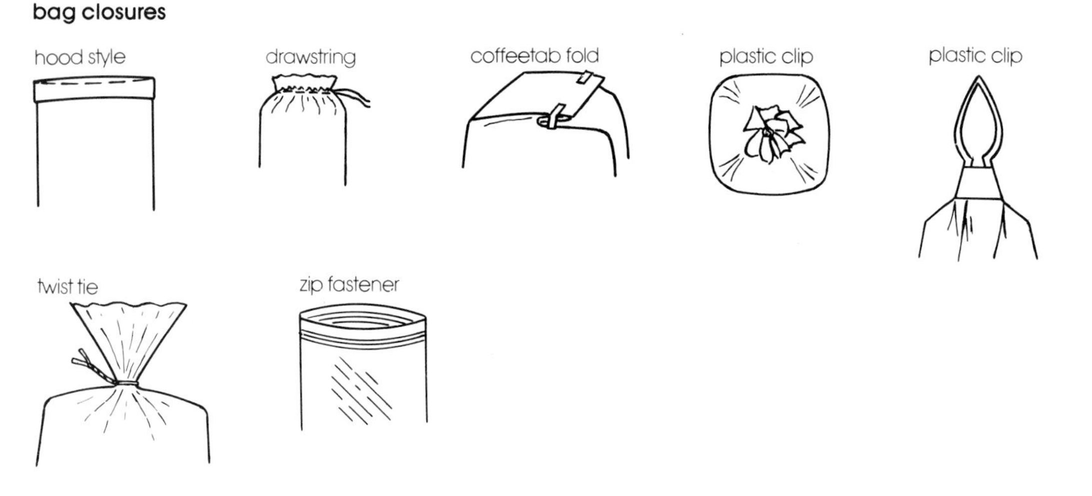

Figure 5-4C. Types and uses of bags and pouches.

high-density polyethylene for industrial products.

The *self-opening style,* or *automatic bag,* has side gussets and a built-in flat bottom. The popular grocery and shopping bags are in this group (Fig. 5-5). The *sachel-bottom* bag has no gusset in the sides and has a flat base. It is used for bulky materials, mostly foods such as coffee and flour.

In addition to these four styles of bags, there are numerous *multiwall bags,* which are used for bulky products, that come in large or small sizes. Large multiwall bags are used for products such as cement, plaster, polymers, and fertilizers. Small multiwall bags are used for coffee, rice, and other bulk food stuffs. Multiwall bags consist of several layers; the number of layers and material used for them depends upon the contents of the bag. For example, a multiwall bag for cement would have a paper outside and plastic liner; for food it would have specially treated and laminated materials.

A variety of closures is available for bags, including sealers, twist ties, plastic clips, coffee tabs, and drawstrings.

Form-Fill-Seal Pouches

The first totally automated package form was the *form-fill-seal pouch.* This package comes in the form of a roll and is filled and sealed on high-speed equipment. Available in a variety of styles and shapes, the pouch has been responsible for the development of many new products. Examples include boil-in-the-pouch frozen foods; instant soups, coffee, and gravies; alcohol and fragrance dabs; instant shoe shines; and packs for pharmaceuticals.

A recent form of flexible packaging, originally developed in Europe, is the

Figure 5-5. The self-opening, automatic grocery bag. *(Photo courtesy of the Great Atlantic & Pacific Tea Company, Inc.)*

tetrahedral (pyramid-shaped) pouch. It is used for fresh milk, since in these containers milk has a shelf-life of six to eight weeks. Pouches for liquids are not new, of course; wineskins were used in biblical times. Wine, motor oil, and soft drinks are also being marketed in flexible packages.

Space-age packaging, originally designed for the Apollo flights, used plastic-laminated pouches. The development of the retort package and aseptic packages (see below) were originally created as a result of the needs of astronauts. Today, these types of packages are used for popular consumer products such as dairy items and fruit juices.

Sterilized snap-open pouches are used for medical kits, surgical gowns and gloves, and other disposable hospital supplies.

Another type of pouch is the *microencapsulation* pocket. It contains the product and also serves as the applicator. Liquid floor polishes and cleaners come in disposable urethane applicator pouches with vinyl-coated paper "hats," which serve as sponge applicators. Shoe polish is marketed in microencapsulation pockets. The polish is trapped between the two layers of a polyethylene-paper pouch and released upon application.

The *bag-in-box* is a practical packaging form that is appearing increasingly frequently on supermarket shelves. It is used for products such as snacks, cereals, and cookies. Different types of paperboards and laminates can be used to create the outer box. The inner bag is made of plastic film, aluminum, or special paper laminates.

The *retortable pouch,* or *retort package,* has been developed to permit sterilization of a product in its package (Figs. 5-6A and B). The pouch with its product can be stored at room temperature and has a shelf life of as long as seven years. The pouch is polyester-foil-polypropylene laminate with specially developed thermal

Figure 5-6A. The retort package. *(Photo courtesy of NASA.)*

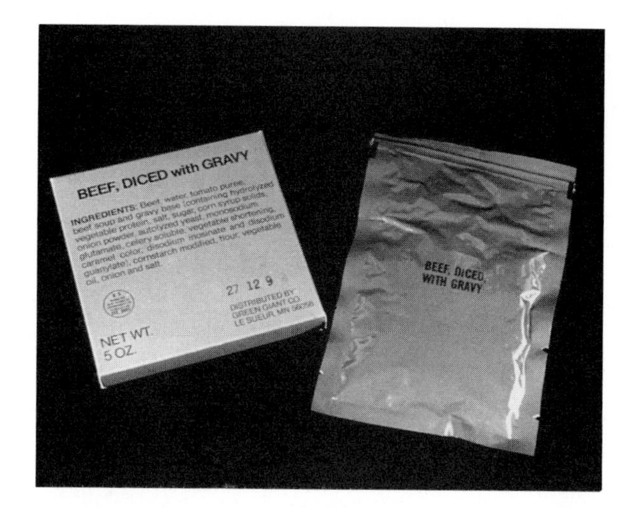

Figure 5-6B. The retort package. *(Photo courtesy of NASA.)*

adhesives. The U.S. Armed Forces has replaced its traditional C-ration cans with lightweight retort pouches.

A variation of the retortable pouch is the *institutional-size pouch.* Because of the costs of packaging material and transportation, this lightweight, low-cost package is replacing the large metal cans that have long been a fixture of the food service industry.

Aseptic Packaging

Aseptic packaging is an exciting new technology in the packaging field. An *aseptic package* is a sterilized container or pouch suitable for food products, beverages, and pharmaceuticals (see Figs. 5-7A and B). Aseptic packages have a long shelf life and do not require refrigeration either during transport or in the store.

Two major aseptic packaging systems are currently available. The first system, *paperboard-based* aseptic packaging, includes the popular brick-style box made from a polyethylene-paper-foil-polyethylene laminate. It is used for juices, wine, tomato sauce, and other liquid foods and beverages. Another type of paperboard-based aseptic package is made from preprinted roll-stock material formed into a tubular shape. The material is a polyethylene-board-polyethylene-aluminum lami-

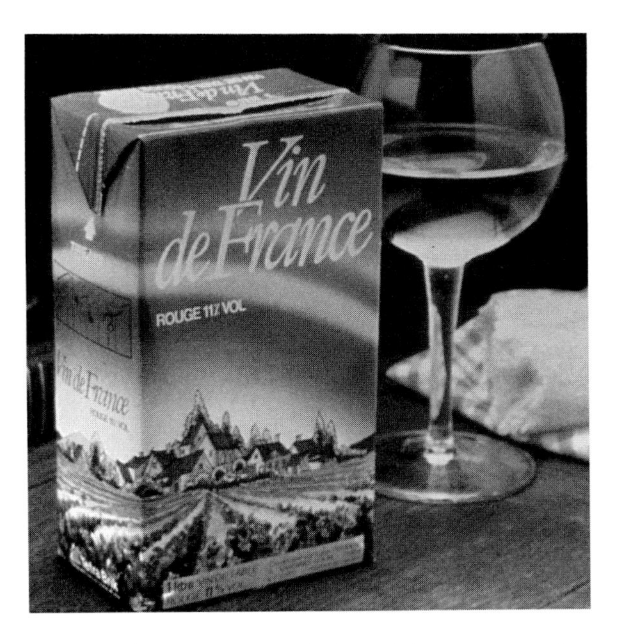

Figure 5-7A. Aseptic packaging. *(Photo courtesy of Brik-Pak, Inc.)*

nate. The lid material (i.e., can ends) is a PVC-aluminum-polyethylene laminate.

The second system of aseptic packaging is *plastic-based.* This type either uses laminated multiple plastics or is coextruded into a single sheet that can be thermoformed into a container. Coextruded plastics for aseptic packaging include ethylene vinyl alcohol (EVOH) and polyvinylidene chloride (Saran). These polymers are sandwiched between layers of adhesive, or "tie layers." The multilayered sheet is a minimum of five layers thick; it is often seven. This package form is used for fruit juices, apple sauce, tomato sauce, and a variety of puddings (Fig. 5-8).

Figure 5-7B. Aseptic packaging. *(Photo courtesy of Brik-Pak, Inc.)*

Figure 5-8. Retort and aseptic packaging. *Left to right:* Igor Bezenyan and Lisa McGowan, designers.

Figure 5-9. Thumbnail sketches are the first step in turning an idea into a finished design.

SHOPPING BAGS

Advertising with Shopping Bags

Shopping bags are the most effective way of advertising a store and in some cases a product or service. They are actually moving billboards. As an advertising medium, shopping bags are used by stores, organizations, banks, churches, labor unions, manufacturers, and even political parties—to name just a few.

Designing Shopping Bags

To design a successful shopping bag, it is necessary to start with a basic idea or concept presented in thumbnail form. Figure 5-9 shows some of the twenty or so concepts a designer originated in three hours in rough thumbnail form. Note the variety of ideas, ranging from simple typography to unusual, sometimes humorous concepts.

When the final concept has been worked out, the design may be transferred to the bag through various media, such as photography collage, photocopy, illustration, or even a photostat reproduction in color or black-and-white. Almost any technique can be used in creating a stunning shopping bag.

The construction of the shopping bag is basically simple: The bag has two side gussets. The artwork can appear on the sides, front, and back of the bag. Although manufacturers can provide patterns for comps, most designers prefer to construct their own bag.

There is no limit to the imagination when it comes to designing a shopping bag. Bloomingdale's, the famous New York department store, for example, features such unique and unusual shopping bags that its name does not even have to appear on the bag (Fig. 5-10). Instead, designers have created a "Bloomingdale's look" that is timely and smart. Bloomingdale's designers often follow the trends in fashion, art, cinema, theater, music, home

Figure 5-10. Bloomingdale's shopping bag. *(Photo courtesy of Bloomingdale's.)*

furnishings, interior design, and architecture. The success of Bloomingdale's shopping bags illustrates how a well-informed, imaginative designer can continually come up with fresh, new ideas.

The design student's favorite project is the shopping bag. Some highly imaginative student-created bags are shown in Figures 5-11A through D.

Figure 5-11B. Shopping bag. F.I.T. class project. *(Photo by Miles David Sebold.)*

Figure 5-11A. Shopping bag. Brenda Miller, designer. *(Photo by Miles David Sebold.)*

Figure 5-11C. Shopping bag. Jean Batthany, designer. *(Photo by Miles David Sebold.)*

PROJECT

1. Create several concepts for a shopping bag. They can be for a commercial (store) bag, a bag with a message, a miniature bag for sales promotions, a bag using only typography, or some other idea. Make thumbnail sketches of your concepts, then construct the bag with side gussets. Create an unusual, exciting, graphically striking design for your shopping bag.

SUGGESTED READING

Packaging Encyclopedia. Des Plaines, IL: Cahners, 1987.

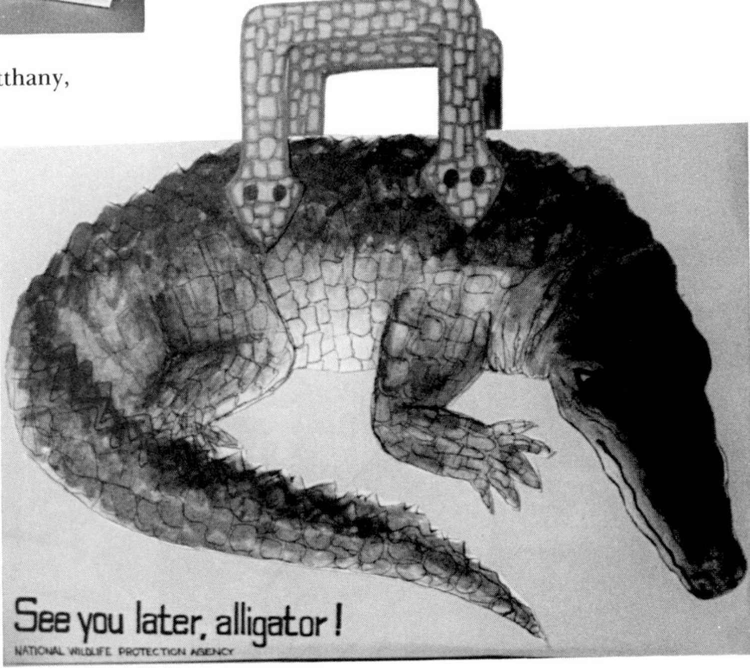

Figure 5-11D. Shopping bag. F.I.T. class project. *(Photo by Miles David Sebold.)*

GLASS CONTAINERS

(Photo courtesy of Parfums Hermès.)

A Short History of Glassmaking / Producing Glass Containers / Types of Glass Containers / Decorating Glass Containers / Packaging Cosmetics / Designing a Fragrance Bottle

GLASS is made from silica (sand), soda-ash, and limestone. In nature it occurs as an opaque substance called obsidian, which is produced by intense heat from lightning or, more commonly, from volcanic eruptions.

No other packaging medium offers the versatility of glass. Glass can be opaque, transparent, clear, tinted, reusable, disposable, and recyclable. It is resistant to acids, most chemicals, and high and low temperatures; it is thermally conductive and electrically insulating.

As a packaging medium, glass is found everywhere, from practical food containers to elegant cosmetic bottles and jars. Drugs and pharmaceutical products are kept pure and safe in glass bottles and vials. Soft drinks look inviting in a clear glass bottle. White, green, and amber wine bottles provide protection from light and preserve delicate flavors.

The popularity of glass containers lies in the nature of glass itself. Glass is chemically inert. That means it does not affect or react to the taste or odor of products packaged in it. Moreover, its smooth, nonporous surface facilitates washing and sterilizing processes, and a glass container can be sealed so that it is airtight.

A SHORT HISTORY OF GLASSMAKING

The earliest known glass objects are beads covered with a green glaze made in Egypt sometime between 10,000 and 3,000 B.C. Since Egyptian craftsmen did not know how to blow or press glass, they poured melted glass into open molds, a slow, laborious process. In time, the Egyptians learned how to make glass containers; the first ones were brilliantly colored opaque bottles and jars.

The Containers

The invention of the blowpipe, which has been dated by some researchers as about 300 B.C., by the Phoenicians, provided the basis for all future methods of producing glass containers. It enabled glass to be blown into molds of all shapes, colors, and sizes, thereby removing glass from the "luxury" list. By 200 B.C., glass manufacture flourished in ancient Rome. Noticing that glass containers were waterproof and did not affect the taste of their contents, merchants began using them to ship wines, oils, and other products on long sea voyages.

Roman craftsmen knew how to make layers of glass of different colors. The famous "Portland Vase" made in Rome about A.D. 70 is an example of this art. The most valuable art glass object in the world—worth about $90 million—it is now in the British Museum.

During the Middle Ages glass was manufactured on the island of Murano near Venice. The Venetians perfected *cristallo* (crystal) glass, the first truly transparent glass. Cristallo could be blown extremely thin and into any shape. As Venice grew in wealth and power, the glassmaking trade flourished and guilds were formed to guard the secrets of the special Venetian craft of cristallo.

By the seventeenth century several other European countries were producing glass. In Bohemia (part of Czechoslovakia) glass blowers produced elaborate, heavy cut glass with many brilliant patterns cut into it giving it a jewel-like appearance.

In the United States, glassmaking was established early in the nation's history. In 1608 a "factory"—actually no more than a hut with a furnace—was producing blown-glass bottles in Jamestown, Virginia. But it wasn't until 1739 that the glassmaking industry was actually established. In that year, Caspar Wistar built a factory in Salem County, New Jersey, to make *flint* (clear) glass. In 1762, another great American glassmaker, Henry William Baron Stiegel, manufactured fine glassware in Mannheim, Pennsylvania.

The famous Sandwich Glass Company was founded in Boston in 1825. It was probably the first glass manufacturer in the United States to produce pressed glass, often called "the poor man's cut glass" since it is an imitation of expensive cut glass. In pressed glass, heavy textured areas are obtained by pressing the glass into molds, often creating a lace-like effect.

The most famous American art glass was made by the Steuben Glass Works, which was founded in 1903 by Frederick Carder at Corning, New York. The Corning Glass Works later acquired the company, and it is now known as Steuben Glass, Inc. The world-famous Museum of Glass at Corning contains more than 15,000 objects made of glass.

By the early nineteenth century, glass production had advanced to using molds in conjunction with blowing to form containers of different sizes and shapes (Fig. 6-1). A "gob" was lowered into the mold and blown to conform to the mold's contours. Somewhat later, glassmakers began to engrave molds to create designs, shapes, and forms. Between 1820 and 1870 manufacturers produced numerous historical or memorial flasks engraved in the likenesses of famous people. Later, glass in the shape of emblems and mottos became popular. Special social and political events were commemorated with glass containers that were made to hold liquor and then to be kept as ornaments. In addition to being decorative, engraved bottles were functional: They provided information about the manufacturer, contents, and directions for use. Most were used for patent medicines and liquor.

The use of glass spread rapidly. In 1858 Thomas Landis Mason patented the famous Mason jar. By 1880 commercial

Figure 6-1. American glass bottles (nineteenth century).

food packers were using glass containers. Beginning in 1899 milk producers abandoned the traditional metal pail for elegant glass bottles designed by Dr. Thatcher of Potsdam, New York.

In 1903 Michael J. Owens of Toledo, Ohio, invented the first semiautomatic bottle-making machine, the most significant advance in glassmaking since the discovery of the blowpipe. By 1904 Owens' machine was completely automatic. It produced containers of equal length and capacity, thereby making uniform glass production a reality.

The Closures

As glass containers gained in popularity, different closures were developed for them. The closures had to be suitable both for the types of containers and for the products they contained.

At first, corks were the most widely used closures. Wine bottles were molded with a ring on the neck enabling the cork to be secured with a string or wire. Most containers for patent medicines, cosmetics, and alcoholic beverages were also sealed in this way. Metal screw tops, introduced with the Mason jar, were quickly adapted to other types of containers and bottles (Fig. 6-2). With the invention of the rubber sealing ring, which guaranteed an airtight closure, other sealing materials such as cardboard and paper were adapted for use on glass containers.

Figure 6-2. A variety of glass jars. *(Photo courtesy of W. Braun Company, Inc.)*

With the introduction of carbonated soda water, the forerunner of bottled soft drinks, it became necessary to invent a closure that could withstand the pressure generated by carbonation. William Painter's invention, a metal cap with a thin cork liner, proved to be airtight. Called the *crown cap*, it was a shallow metal disk with a flared, fluted "skirt" that was forced against the ring on the bottle neck to create a tight seal. The cap could easily be removed with another new invention—the bottle opener. Lug caps, vacuum caps, and plastic or metal friction-fit caps are all outgrowths of the development of closures for containers.

The advent of plastics added a new dimension to closure development (Fig. 6-3). Thermoset plastics, ureas, and phenolics were used in the early twentieth century. They were favored for their rigidity and their resistance to chemicals—properties that made them suitable for use with many chemical and pharmaceutical products.

Less expensive closures than those made from thermosets and ones that could can be recycled are those made from thermoplastics. Thermoplastic materials used in closures include polyethylene, polystyrene, polypropylene, and SN and ABS copolymers. Thermoplastic clo-

Figure 6-3. Plastic closures, applicators, and pumps. *(Photo courtesy of W. Braun Company, Inc.)*

sures can be designed with dispensers, such as flip-tops, spouts, and other pouring devices.

PRODUCING GLASS CONTAINERS

Production Process of Glass

Four basic processes are used in producing glass: blowing, drawing, pressing, and casting.

Blowing uses compressed air to form molten glass in the cavity of a mold. Most commercial glass bottles are produced on completely automatic equipment using this method.

Drawing is a process in which molten glass is pulled through dies or rollers that shape the soft glass. Tubes, sheet, and plate glass are produced in this way.

Pressing is a process in which molten glass is pressed against the sides of mold with a plunger. Lenses, paperweights, baking dishes, tumblers, and insulators are among the many items produced by pressing glass.

Casting uses gravity or centrifugal force to cause a gob of molten glass to form in the cavity of the mold. Casting is used to produce art glass and glass for architectural and decorative use.

The Finish

The portion of the bottle that contains the opening and accommodates the closure is termed the *finish*. There are three basic types of finishes: thread, lug, and friction. *Thread finish*, the most common closure and finish, uses a metal or plastic screw cap. The threads of the cap engage the threads molded on the neck of the bottle. In a *lug finish*, sometimes called vacuum finish, a cap is forced by pressure onto the neck of a container. *Friction finish*, which is similar to lug finish, works by friction over the opening. It is a very tight finish.

There are also some special types of finishes, including sprinkler tops, roll-ons, snap caps, and pour-outs. *Sprinkler tops* are used for dispensing cosmetics, household cleaners, and pharmaceuticals. *Roll-ons* are similar to ball point pen tips. They are used for deodorant containers and applicators. *Snap caps* (usually an overcap) are similar to friction finishes. They are used for aerosol cans, cosmetics, and pharmaceutical containers. In addition, many variations exist on these finishes.

In order to protect the public, especially children, in recent years a wide variety of safety finishes have been developed. These tamper-proof and childproof systems, as they are known, include press and turn, press and lift, and the combination lock (turn the cap and line up the dots or arrows). They are used mainly for drugs and other health and pharmaceutical products and are mandatory on some product containers.

In the interest of economy and efficiency, glass finishes have been standardized and are designated by number. The standards are on file at the Glass Container Manufacturers Institute (GCMI) in

Washington D.C. Glass finishes are designated by the size, in millimeters, of the outer diameter of the bottle. For example, GCMI finish 400 (the most widely used type of finish) is a shallow continuous thread.

TYPES OF GLASS CONTAINERS

Structurally, glass containers and bottles have changed significantly in the last 35 years. Through improvements in design and materials—most glass containers have shed about 55 percent of their weight—larger containers have become feasible. Today there are many basic types of glass containers in use; some are described here.

Bottles are the most extensively used glass containers. They may be cylindrical, oblong, or rectangular. The neck is almost always round to permit easy pouring and effective closure. Today, well over half the glass bottles produced in the United States are used for beverages, such as wine, beer, liquor, and soft drinks (Figs. 6-4 and 6-5).

Jars are simply wide-mouthed bottles. The large opening accommodates utensils and fingers. Certain types and shapes of jars are suited to particular kinds of products, primarily cosmetics and food. Jars usually have a low center of gravity, making them easy to store and transport.

A *tumbler* is similar to a jar. It is shaped like a drinking glass and used in packaging foods such as jams, jellies, spreads, and sauces.

Figure 6-4. Wine bottles. Label designs by Rhonda Cericola. *(Photo by Miles David Sebold.)*

A *jug* is a large bottle with a short neck and a carrying handle. Jugs are used for water and wine.

Carboys are heavy-duty industrial shipping containers used mainly for chemicals. Their capacities range from three to thirteen gallons, and they are transported in wooden or plastic crates.

Vials are small, flat-bottomed glass containers. They are tubular in shape and may have any of a variety of neck finishes. Vials are widely used for antibiotics and other pharmaceuticals. Plastic clo-

Figure 6-5. Beer bottle models. Label designs by Christine Gangi, Tom Li, and Diane Jones.

sures and tamper-proof devices are often used on these strong containers.

Ampoules made from glass tubing are used for serums and injectable drugs. After the product has been poured into the ampoule, the open end is melted and sealed shut. The ampoule must be broken at a designated breakline before the contents can be used.

DECORATING GLASS CONTAINERS

Decorated glass containers are both functional and attractive. An often used form of decoration is the *label*, which can be die cut in various shapes and adhered to the bottle. Types of labels include front, back, full wrap, and neck bands. Pressure-sensitive labels and decals are also used on glass containers, especially those for cosmetics. Pressure sensitive labels, which can be peeled off, are used for removable price labels. Decals are transfers that create the impression of printing or screening on glass.

Another frequently used method for decorating glass is silkscreening on the glass in metallic and ceramic colors. Other special effects, such as frosting (creating a cool, delicate effect with acids), hot stamping (a mechanical transfer method done with metallic or nonmetallic colors to create a delicate look for containers; used mostly on plastic bottles), and transparent colors, are also possible.

Thermoset powders are another innovation in glass decorating. The powders can be cured at relatively low temperatures to provide an overall finish in an almost unlimited range of colors.

The ways in which to decorate closures are equally numerous. These include printing, screening, embossing, hot stamping, or vacuum metalizing.

PACKAGING COSMETICS

The properties of glass, as well as its elegance and beauty, make it ideal for packaging cosmetics. Cosmetics, like drugs or pharmaceuticals, require chemically inert containers to ensure a long shelf life. Glass has that property.

The cosmetics industry produces products in a broad price range, and the bottles in which the products are packaged closely reflect that range. A fine fragrance or lotion is an exclusive, luxurious substance, and the superbly crafted bottle, jar, or aerosol that contains it serves as a visual symbol of the product.

DESIGNING A FRAGRANCE BOTTLE

There are fashions in bottle designs, as there are in clothing. The fashion ranges from refined simplicity to complex shapes and combinations. In recent years, for example, a long, exaggerated closure over the bottle has been popular. Years ago, round and flared closures were fashionable. To be able to interpret trends in design, designers must keep informed in all fashion areas.

A good designer must work productively with technical, marketing, advertising, and production personnel. This may entail exercising restraint and self-discipline. Before developing conceptual sketches, the designer must consider all facets of the project: the product, marketing areas and needs, the target consumer, the retail price, possible companion items (e.g., bath products, soaps, bath oil, or dusting powder), and TV and print advertising. Only when all of these elements are clearly understood can the design program begin.

Basic Approaches to Designing a Bottle

There are two basic approaches to designing a fragrance bottle or container (or any bottle or container): either using an existing bottle, called a *stock bottle*, with the appropriate closure, or creating an original concept known as a *private mold*.

Each procedure has its merits. The stock bottle is a wise choice when cost and time are key factors. A wide selection of stock bottles and closures are available at standard prices in flint, opal, and even color. Competent designers recognize that the appropriate stock bottle offers plenty of opportunity for imaginative design. Be aware, however, that the term "stock bottle" does not necessarily mean that they are readily or immediately available. Some stock bottles are manufactured in minimum quantities at infrequent intervals.

The private mold is suitable when a new fragrance (or product) is introduced (see "Using Private Molds" below). It must be unique since its design has to be recognized by consumers for its unusual shape, color, or decoration. Eventually, a private mold will become a visual symbol of the product as well as the corporation or manufacturer.

Using Stock Bottles

The ideal stock bottle is simple, well designed, and pleasing in its contours and proportions. Combining the right stock bottle and closure with a distinctive label or decorative effect can result in the perfect container for a special fragrance (Fig. 6-6).

An advantage of the stock bottle is that the manufacturer or distributor can supply samples that can be used to create a wide variety of designs. This avoids the necessity of making sketches, mock-ups,

Figure 6-6. A selection of stock glass bottles. *(Photo courtesy of W. Braun Company, Inc.)*

and expensive three-dimensional models. Some stock houses publish heavily illustrated catalogs of more than 1000 items to be used by designers in creating attractive packages.

Using Private Molds

When designing with a private-mold bottle, a tasteful design, using sophisticated decorative techniques, can convey feelings of richness, elegance, and exclusiveness. The designer aims for a classic, timeless look and avoids clever novelties. The name of the fragrance is incorporated into the design theme, and the fragrance and bottle are treated as equal partners in a work of art. Incorporating the name of the fragrance into the design of the bottle contributes to the success of the product. The bottle then becomes a visual symbol of the product, easily recognized and identified.

Phases of Designing a Bottle

Like all design procedures, the design of a fragrance bottle should be carried out in three phases:

1. Preliminary design concept
2. Comprehensives or three-dimensional models.
3. Production models, blueprints, and mechanicals.

Phase 1: Preliminary Design Concept

The preliminary design concepts are presented to management and marketing and advertising personnel. At the design meeting, the designer must exhibit integrity and conviction, but above all he or she must exercise tact.

The design process may require a lot of roughs—the more, the better (Figs. 6-7 and 6-8). Present at least ten variations, but have all of your roughs and thumbnails available. Someone is sure to ask for more ideas.

Never discard thumbnails or rough sketches; keep them all together in a folder. You'll need to have them on hand until the final design has been chosen, and they are useful for future reference.

Phase 2: Three-Dimensional Models

When a final design is chosen, phase 2 begins. This phase consists mainly of preparing a three-dimensional model.

First, the designer creates a mock-up from plasticene (Fig. 6-9, left). Since it is easy to change the details on plasticene models, plasticene is an excellent medium for preparing variations of the proposed bottle and studying and evaluating various aspects of the design, such as the proportions of the bottle, the surface for labeling or decorating, and the closure. It is advisable to present this model to the

Figure 6-7. Sketches for cosmetic bottles. *(Photo courtesy of W. Braun Company, Inc.)*

Figure 6-8. Marker sketch for fragrance bottle by Eliana Themistocleous.

manufacturer for comments and suggestions before having a costly lucite model prepared by a professional model maker. (Note that the left-hand bottle in Fig. 6-9 is a polyester model.)

The lucite model (Figs. 6-10 and 6-11) is an exact replica of the proposed bottle, complete with a removable closure (or closures), precise in every detail, calibrated by computer to the exact capacity required, and beautiful to look at. Occasionally, several lucite models are prepared to be used in making a final selection or for doing market testing or experiments.

Figure 6-9. Fragrance bottle models by Rhonda Cericola. *(Photo by Miles David Sebold.)*

Figure 6-10 (above). Lucite fragrance bottle models, prepared by expert model maker. *(Design and photo by the author.)*

Figure 6-11 (right). Plexiglass and lucite model for a fragrance bottle. *(Photo courtesy of Helena Rubinstein, Inc.)*

Phase 3: Production Models, Blueprints, and Mechanicals

When the final version of the bottle has been selected, phase 3 begins. In this phase the bottle manufacturer takes over. Blueprints and some additional prototypes are prepared for the purpose of making a mold. This is a highly specialized, time-consuming job. It may take weeks or even months to prepare the mold that will produce an original, unique bottle.

Figure 6-12. Fragrance bottles. *(Photo courtesy of Chanel, Inc.)*

World-Famous Fragrance Bottles

Figures 6-12 through 6-18 show some of the world's most famous fragrance bottles, ones that over time have come to be viewed as classics.

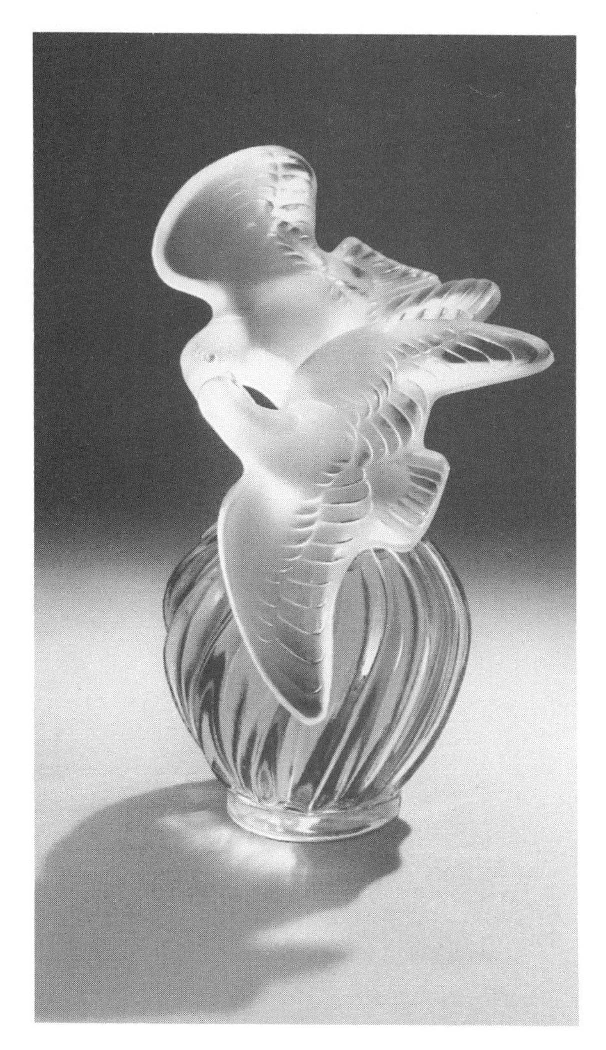

Figure 6-13. Fragrance bottle. *(Photo courtesy of Parfums Nina Ricci.)*

Figure 6-14. Fragrance bottle. *(Photo courtesy of Parfums Hermès.)*

Figure 6-15. Fragrance bottles. *(Photo courtesy of Jean D'Albert.)*

Figure 6-16. Fragrance bottle. *(Photo courtesy of Charles of the Ritz Group, Inc.)*

Figure 6-17. Fragrance bottle. *(Photo courtesy of Houbigant, Inc.)*

Figure 6-18. Fragrance bottles. *(Photo courtesy of Calvin Klein Cosmetics.)*

PROJECTS

1. Using a stock bottle, come up with a design for a cosmetic product of your choice. Be sure to choose an appropriate closure, and design a distinctive label or decorative effect.

2. Develop some concepts for a private-mold bottle to be used for a costly fragrance. Construct a three-dimensional model of your favorite concept from plasticene and refine your design.

SUGGESTED READINGS

Glass Packaging (trade publication).

Helen and George McKearin. *Two Hundred Years of American Blown Glass.* New York: Crown, 1946.

TRADE ASSOCIATIONS

Glass Packaging Institute, Washington, D.C.

Society of Glass Decoration, Port Jefferson, N.Y.

COLLECTIONS

Metropolitan Museum of Art, New York, N.Y.

Corning Glass Works, Corning, N.Y.

7

CANS, TUBES, AND AEROSOLS

(Comprehensives designed by Anne Walker and Rhonda Cericola. Photo by Miles David Sebold.)

Designing Cans / Metal Tubes / Plastic Tubes / The Aerosol Can

NAPOLEON Bonaparte could be called the godfather of the metal can. In 1795 he authorized a competition in which 12,000 francs would be awarded to anyone who could devise an effective way to preserve food. Nicolas Appert developed a method for putting partially cooked food into glass containers with cork closures and then immersing the containers in boiling water. He won the prize and subsequently published a paper on the use of heat and evacuation of air to preserve food.

In 1810 Peter Durand received a patent for the technique of putting foods in "vessels of tin, glass, pottery, and other metals." Since glass is fragile, Durand proposed the use of cylindrical canisters made of iron with a tinplate coating.

The first commercial cannery was established in 1813 by two other Englishmen, Bryan Donkin and John Hall. It supplied canned foods to the British army and navy. The early metal can from this first cannery was handmade, with the side seam and end sections soldered. Food was inserted through a hole on the top, and a small disk was soldered over the hole. The disk had a tiny hole in it to permit the escape of air that had been heated during the filling process. Soldering this hole completed the canning process.

By 1819 the can had come into use in the United States. A New Yorker, Thomas Kensett, and a Bostonian, William Lyman Underwood, were already canning foods in glass containers. But it was Kensett's 1825 patent that initiated the development of the tin can. By 1847, a process had been devised for forming the flanged ends mechanically. The Civil War stimulated further development of the can. In 1867, side seaming became a mechanical process. By the turn of the century, there were more than 1800 canneries in the United States.

The most important development of the early 1900s was the open top cylindrical can, which would be the standard for years to come. Over the years, oval, oblong, square, rectangular, and fluted cans would be made for products such as paints, food, tobacco, and industrial products (Figs. 7-1 and 7-2). Spray cans were introduced in the early 1940s.

The metal can has changed greatly in the twentieth century. Today's cans are made of tin-free steel lined with various plastic resins, such as acrylics, vinyl,

Figure 7-1. Some early American cans (1920). *(Photo courtesy of the Container Corporation of America, Packaging Museum.)*

epoxy, and alkyds (Fig. 7-3). The introduction of tin-free steel made entirely new concepts in can making possible. It allowed, for example, the side seam to be held together with organic cement or by welding rather than by soldering. The absence of solder lets designers decorate the entire body of the can. Figure 7-4 illustrates the can-making process used today.

Because of aluminum's light weight and resistance to corrosion, it is used for cans that contain beer and carbonated soft

Figure 7-2. Oatmeal canister (1912). *(Photo courtesy of Landor Associates.)*

drinks. Moreover, two-piece, drawn aluminum containers have no side seam and can be printed around the entire body.

Composite cans and containers, which are used for frozen fruit juices and premixed dough products, are actually spirally wound fiber (paper) containers (see, for example, the can pictured in Fig. 7-5). Although paper is the primary component of composite can bodies, several layers of foil or plastics are often used. Molded plastic cans are sometimes used for soft drinks.

Aluminum foil containers were originally developed to supply bakers with disposable baking pans and pie plates. Foil containers are used today for frozen foods, cereal products, school food service products, and pharmaceuticals.

Figure 7-3. U.S. Army "C" rations (1942).

comparative side seams

soldered side seam

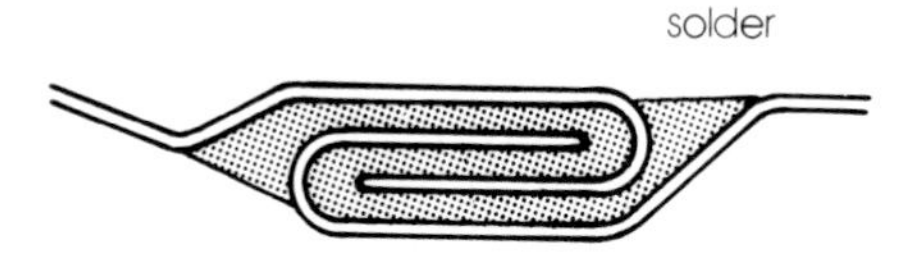

cemented side seam

welded side seam

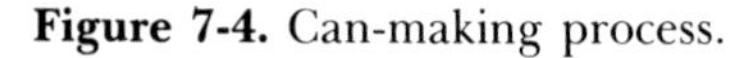

Figure 7-4. Can-making process.

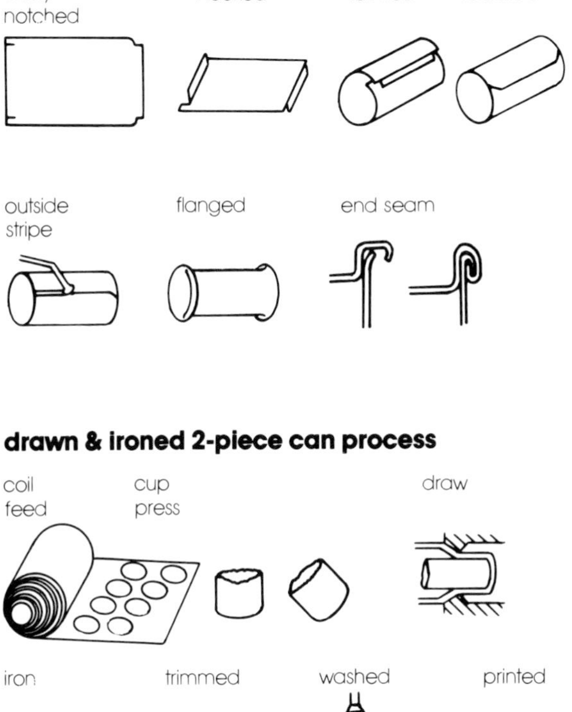

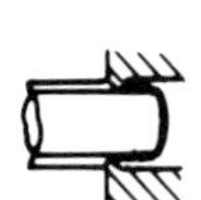

conoweld welded can process

body blank feed

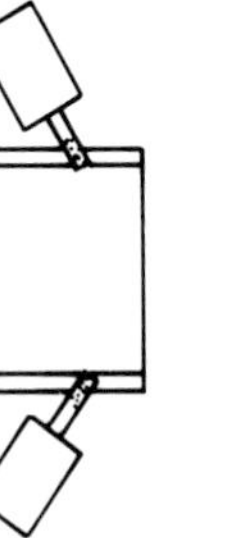

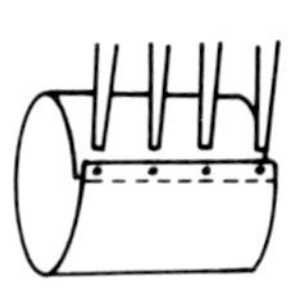

seam welding

miraseam cemented can process

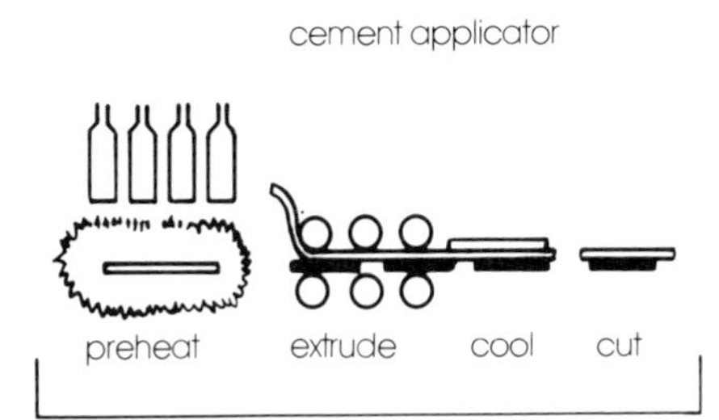

body maker

notch and trim · heat · form cool bump (2nd cool bump)

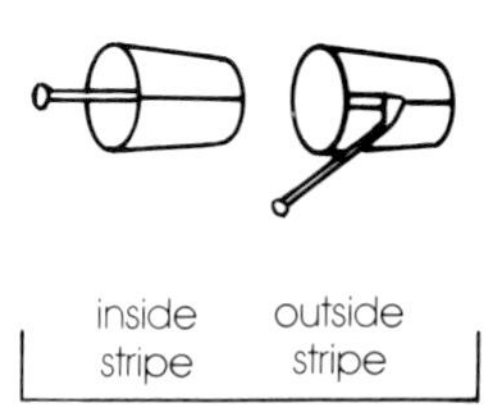

Figure 7-4. (*continued*).

Figure 7-5. Children's bath product packaging by Ronit Spitzer.

DESIGNING CANS

Cans have generally had labels, but the development of new printing techniques has given rise to using them even more as a selling device. It is important for the designer to be aware of these developments—both in printing and in advertising—when planning the graphics for a container.

For many years the paper label has been an important part of the soldered metal can. In fact, a large percentage of cylindrical metal cans, especially food cans, still use paper labels. But modern high-speed printing processes now permit lithographic printing of graphic illustrations or photographs—in as many as six colors—directly on the can. A raised surface (embossing) adds sparkle, and the use of contrasting, transparent colors enhances the appeal and graphic impact of the shiny metal container.

The Brand Name

The most important aspect of the design for a can is the product's *brand name.* Package identification by brand name is a study in itself. The brand name is the verbal part of the trademark and can be represented in either symbolic or pictorial form (Fig. 7-6).

Because the can is a round object, visibility and readability of the brand name may be a problem. Designers must therefore keep in mind that a flat design appears totally different when it is wrapped around a round object. This is basically an optical problem and must be treated and solved as such.

Thus, a brand name on a round package can be successful only when it is visi-

Figure 7-6. Examples of contemporary cans. *(Photo courtesy of the Great Atlantic & Pacific Tea Company, Inc.)*

ble, appealing to the eye, and suggests value and quality. Many popular soft-drink cans are designed with this important principle in mind. The Coca-Cola, 7 Up, Pepsi Cola, and Sunkist cans are especially striking.

The Label

Another component of a can's design is the *label*. The illustration or photo on the paper label must serve as a "transparent" container. That is, it must enable the consumer to "see" what's in the can (Fig. 7-7). Thus, along with type or a hand-lettered logo for the brand name, there

Figure 7-7. Comprehensives for soft drink cans. Tami Joznick and Sonia Biancalani, designers. *(Photo by Miles David Sebold.)*

can be a striking product illustration. Visit your supermarket to get some ideas from the canned foods and soft drinks displayed there.

The Designing Process

To design a successful can, designers must be familiar with the product it contains. Imagine you are going to design a container for a new soft drink called Cool, which comes in an aluminum can. The design will be printed directly on the can, so you can take advantage of the shiny reflective surface and print with transparent colors.

The first thing you need is a logo for the name Cool. But since the can is a round (not flat) surface, you have to solve the optical problem of lettering on a round can before you can develop your design concept for the logo.

Start with an actual can. Wrap paper around it, set the can on a shelf or table, and step back to look at it. Determine the area your eyes can see without turning the can. This should give you an idea of how much space you need for the logo, whether you should use a vertical or slanted logo, and whether you should use symbols, an illustration, or some other device to give your product "shelf appeal." All these details must be worked out in a rough form *directly on the can.*

After you come up with three or four good design possibilities, render them on paper, acetate, or foil, using transparent inks or markers. Then wrap your sketches around the can, thereby creating your comps. If you work on acetate, be sure that the design will face the can (i.e., paint the design in reverse). This way the paint will not smear. You can use cut-out lettering from transparent, self-adhesive films, such as Color Key or INT, to create clearer, more concise comps.

Once a final design has been chosen, the next step is to do a mechanical, usually specified by the client's production department. The can manufacturer will provide the size, shape, and die sheet for the mechanical. Do not proceed with the mechanical without this important information. In addition, always consult your supplier if something is not clear to you. In that way, serious errors can be avoided. Above all, do not accept sizes on the telephone; they should be in writing on the die sheet.

METAL TUBES

The first collapsible *metal tube* was developed in 1841 by an unknown artist seeking a more efficient way to dispense oil paints. In the 1890s a dentist, Worthington Sheffield, invented a new dental cleaner—toothpaste—and packaged it in a collapsible metal tube.

Tubes are lightweight and provide excellent product protection. Today more than half of the metal tubes manufactured are used to dispense toothpaste. They also serve as safe, sanitary dispensers for pharmaceutical and cosmetic products.

Producing Metal Tubes

Tubes are produced from stamped or punched metal slugs. The slugs are fed into extrusion presses, and the resulting tubes are conveyed to a machine that trims them for crimping and threads the neck ends for the appropriate closure. The tubes are then conveyed to an offset printing machine that can print in up to four colors. After drying, the tubes are automatically capped; they are filled from the bottom and then sealed and crimped.

Materials Used for Metal Tubes

There are three basic tube materials:

1. *Aluminum tubes* are used for toothpaste, shaving cream, toiletries, and food.

2. *Tin tubes* are made from real tin and are strong, durable, and chemically inert. They are used when compatibility is critical, such as for dispensing medicinal-pharmaceutical ointments.

3. *Lead tubes* are less expensive than aluminum or tin tubes. They can be lined with various coatings, depending on the substance they will contain. Paints, adhesives, and chemicals are among the products packaged in lined lead tubes.

Designing Metal Tubes

The most important consideration in designing a metal tube is compatibility between the package and the product. The type of metal used and the lining must be appropriate for the product the tube will contain. The decision about what materials you choose may require information provided by a chemist or the packaging engineer.

PLASTIC TUBES

Plastic tubes, which are available in stock sizes or can be custom made, are unbreakable, leakproof, and inexpensive. They can be either transparent or opaque in many colors.

Plastic tubes are made from extruded polyethylene, polypropylene, or laminated plastics and can be lined internally. The tubes are sealed either by applying heat or ultrasonically.

Plastic tubes are used to pack water-based and oil-water emulsions, such as shampoos, lotions, and hair-grooming products. They have recently come into use for food products, including tomato sauce, cheese spreads, and dairy products.

Designing Plastic Tubes

Basically, the design process for plastic tubes is the same as that for metal tubes—except there are greater variations for the decorations on plastic tubes. Decorative processes for plastic tubes include offset printing, silkscreening, hot stamping, and embossing. To prepare a mechanical for a plastic tube, the designer needs to obtain the proper layout from the manufacturer. For comps, the manufacturer will provide blank tubes in a variety of sizes and colors and with different types of closures.

THE AEROSOL CAN

The modern *aerosol can* is a result of research in the early 1940s by Lyle D. Goodhue and William N. Sullivan. A metal container was pressurized with gas and fitted with a push-button dispenser to deliver a spray of insecticide. This invention, called the "Bug Bomb," was used by the U.S. Armed Forces (Fig. 7-8).

Figure 7-8. The modern aerosol dispenser derives from the "Bug Bomb" used during World War II.

Figure 7-9. Aerosol can comprehensives. *(F.I.T. packaging design class project.)*

After the war the aerosol was converted to peacetime applications. It was used to dispense hair sprays, fragrances, toiletries, foods, medications, paints, and many other consumer goods and industrial products (Fig. 7-9). It eventually became known as the pressure packaging system.

Operating Aerosol Cans

An aerosol can is an airtight, valved container made from metal, glass, or plastic and filled with a formulation that consists of a propellant (a gas) and the active ingredient (the product). When the valve is operated, the pressure of the gas pushes both the propellant and the active ingredient through a small pinhole opening in the valve. A fine, misty spray is produced; the fineness of the spray depends on the nature of the product and propellant and the type of valve used.

There was evidence that the chemicals (chlorofluorocarbons) used in aerosol cans were eroding the protective ozone layer of the atmosphere, which filters out ultraviolet light from the sun, and could therefore cause cancer and damage to plants and animals. Today, safe propellants are available for all types of aerosols, and new dispensing systems have become available in Europe. In the United States, hydrocarbons (propanes, butane, isobutane) and, more recently, du Pont's Dymel™ propellant are in widespread use replacing the old chlorofluorocarbons. These new propellants and dramatic package designs promise to make the aerosol packaging system even more popular in the future.

Designing Aerosol Containers

Designing aerosols and aerosol containers can be an exciting graphic experience. The metal can aerosol usually has a plastic cap, which can be painted, labeled, and molded in any color. The can itself can be painted in transparent or opaque colors, much like any other metal can. It can also be labeled.

Plastic and especially glass aerosols can be made to look more elegant and therefore more expensive. Glass aerosols should be treated and decorated like glass bottles (see Fig. 7-10). For cosmetic products (e.g., fragrances), the glass can be frosted and shiny gold metallic paint used for lettering and other graphic decorations. The cap can be metalized and custom designed, although stock closures are available.

Figure 7-10. Glass aerosol. *(Photo courtesy of Prince Matchiabelli, Inc.)*

PROJECTS

1. Design a plastic tube that can stand on its closure.

2. Develop a tube for a product that needs an applicator closure.

3. Originate a new product; for example, a new type of foam bath for children, sun tan oil, detergent, hand lotion, spot remover, cake decorator with a different type of nozzle, or antiseptic spray, that uses an aerosol can.

SUGGESTED READINGS

Aerosol Age (trade publication).
Packaging Encyclopedia. Des Plaines, IL: Cahners, 1987.

8

ENVIRONMENTAL IMPLICATIONS OF PACKAGING

(Photo by Miles David Sebold.)

Solid Waste Disposal / Packaging and the Law

OUR early ancestors tossed bones, broken pottery, and other useless rubbish into heaps near their caves. When the heaps of refuse grew too large, the people just moved on; but later, after the development of cities, people could no longer move away from their trash. Thus, in ancient Rome Pliny complained that "the air is foul," and in medieval times streets were strewn with garbage even though there were strict laws against littering (Fig. 8-1).

Figure 8-1. Royal proclamation against pollution. Fourteenth-century print (detail).

Figure 8-2. Garbage bags. *(Photo by Miles David Sebold.)*

Today we still throw away most of the things we use (Fig. 8-2). In fact, we are producing waste in ever-increasing volumes. But it is only recently that we have become aware of some of the consequences of this vast accumulation of trash. Among these consequences are shrinking forests, polluted air and water, and dying animals. As a result of this new awareness, legislation dealing with solid waste disposal has been enacted, and the disposal of packaging materials has become the subject of intense debate. In this chapter we will examine some of the implications of these issues, especially as they relate to the design of packaging materials.

SOLID WASTE DISPOSAL

About 115 million tons of industrial solid waste were created in 1985, with plastic contributing 0.8 percent of the total. The Environmental Protection Agency (EPA) estimates that the amount of industrial waste is growing by about 3 percent per year.

Most municipalities dispose of solid waste by burying it in landfills. But according to the Worldwatch Institute, an environmental research organization, roughly half the cities in the United States will run out of landfill space before 1990. In other words, solid waste disposal in the United States is developing into a crisis situation. As a partial response to this ever-worsening situation, the federal government passed the Solid Waste Disposal Act of 1965 and the Resource Recovery Act of 1970. Both acts made federal funds available for local pilot projects (see Packaging and the Law in this chapter).

Problems from Plastic Waste

The volume of plastic production is almost beyond belief. More than 20 million tons of plastic a year is produced in the United States alone. Indeed, the cubic volume of plastics has surpassed that of steel, copper, and aluminum combined. Abroad, even the least-developed nations are likely to have a plastic manufacturing plant churning out polyethylene and other plastics.

From the standpoint of disposal, the problem is that plastic objects do not rust, dissolve, or evaporate. The plastic materials currently in use will take up to five centuries to degrade.

A related problem is the hazard to wildlife caused by plastics. The National Academy of Sciences, for example, has estimated that commercial fishing fleets dump at least 52 million pounds of indestructible plastic fishing lines and nets into the oceans each year. The resulting annual death toll, experts say, is about 2 million sea birds and 100,000 sea mammals (whales, seal). Intestinal blockages caused by plastic bags are another cause of animal death. The Marine Sciences Research Center at Stony Brook, New York, has estimated that 30 percent of the fish in the world's oceans have pieces of plastic in their stomachs that interfere with digestion.

Using Biodegradable Plastic

In principle, chemists know how to make biodegradable plastics, but the materials are relatively expensive to produce. As of 1987, a typical biodegradable plastic costs about $2.00 a pound, whereas ordinary polyethylene costs only 50¢ a pound. Manufacturers do not always have sufficient incentive to produce the less dangerous materials.

About 30 states, however, now *require* that the "six-ring" carriers used for canned and bottled drinks be made from biodegradable plastic. Otherwise, when disposed of carelessly, these carriers become deadly snares for birds and small animals. Biodegradable plastics, on the other hand, will begin to crumble after three or four weeks of exposure to sunlight, greatly reducing the risk to wildlife.

Recycling

To use *recycling* as a method of "disposal," the material must be capable of being recycled and it must have value both in regard to profitability and to second use. Plastic packages can be recycled. Nevertheless, as of 1988, only 10 percent of plastic packaging underwent that process. The reason is that for the vast majority of the plastics manufactured today, the low quality of the resulting material from the recycling process makes the process unprofitable.

High-performance thermoplastics, however, can be reused many times. A quality plastic toy can be recycled into material for an automobile bumper, and the bumper can be recycled into material for office building doors. Moreover, the returnable, refillable bottles made of Lexan resin are being used as many as fifty times; and then they are reground, reprocessed, and remolded to make other products.

Compatibility, quality, and high performance are the most important criteria in the recycling of thermoplastics. Provided that these criteria are met, recycling could be used to reduce the accumulation of solid waste. Theoretically, as long as the recycled plastics are of the same quality as the original plastic, a billion pounds of recycled high-quality thermoplastics can replace a billion pounds of high-performance ABS or PET plastics. Recycling the plastics used in packaging should therefore be viewed as an opportunity to increase the profitability of the packaging industry.

Incineration

The main products of incineration (i.e., combustion) of plastics are carbon dioxide (CO_2), carbon monoxide (CO), and water (H_2O). These products do not cause serious air pollution, provided properly operated and constructed incinerators are used for the disposal of plastic wastes.

In Japan, modern incineration plants that cause minimal pollution and generate energy along with the heat produced are already in operation. The heat energy recovered from the incineration of plastics can be offered to consumers on a daily basis. Some European countries, including Switzerland, France, Germany, the Scandinavian countries, and the Netherlands, also successfully incinerate plastics and other refuse.

Unfortunately, most of the incinerators in use in the United States are obsolete

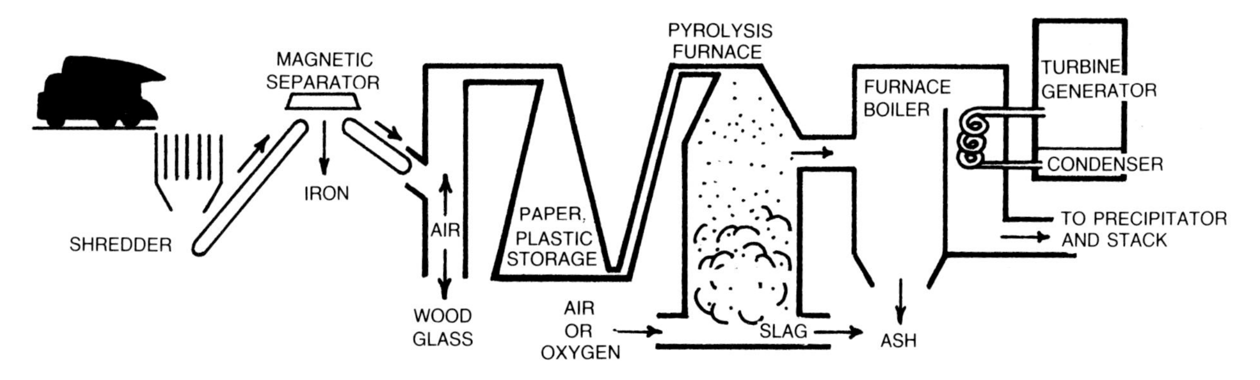

Figure 8-3. Pyrolysis. *(Diagram by Elizabeth Downey. Courtesy of White, Weld & Co.)*

under the standards of the Bureau of Solid Waste Management. These poorly functioning incinerators have actually caused an increase in air pollution in recent years. But several new incineration methods are under study. Among them is the *pyrolysis* system, which burns plastics and other refuse in an oxygen-free atmosphere (see Fig. 8-3). In this system, waste materials can be recovered and reused as commercial chemicals. Model pyrolysis systems are in operation or being tested in several municipalities.

Special Incinerators

The incineration of PVC and PVDC may give off small amounts of hydrogen chloride (HCl), causing some scientists to consider those plastics the most troublesome polymers from the standpoint of incineration. But several types of incinerators have been developed for disposing wastes containing even these chlorine-containing plastics.

Takuma, a Japanese manufacturer, has developed a system that does not produce smoke or emit HCl. Plastic wastes are pulverized and then baked in a kiln. The HCl gas generated by this process is made to react with ammonia and recovered in the form of ammonium chloride, which can be used as an industrial chemical.

Katayose Kogyo Kiki, another Japanese manufacturer, has introduced a different system for incinerating PVC without causing significant air pollution. Loads are moved automatically from one furnace chamber to another, and blowers force gases through the waste materials as they are being processed. The gases dissipate themselves by dissolving and therefore cause only minimal air pollution.

Union Carbide has developed a system for processing liquid or gaseous wastes. HCl is injected into a vortex burner and quenched in a graphite tower. The resulting vapors can be processed to produce industrial chemicals.

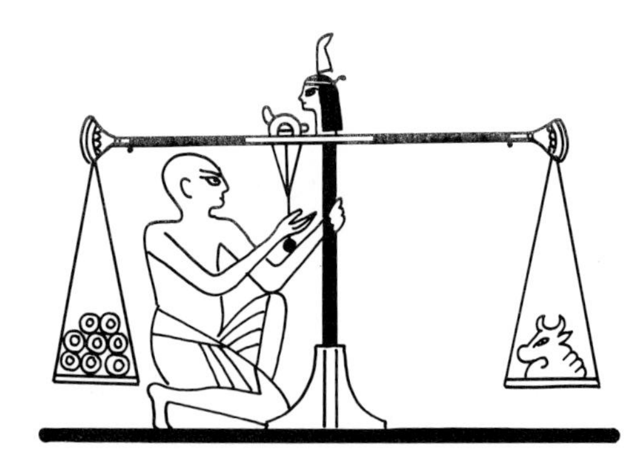

Figure 8-4. Egyptian tomb painting shows government inspector weighing and measuring consumer goods.

PACKAGING AND THE LAW

Throughout this book we have noted that packaging was among the earliest human inventions. Baskets, pottery, and skins were used in prehistoric times to transport and preserve food. Then when people began living in communities and cities and industry, commerce, and trade began to develop, the need for different types of packaging increased accordingly.

Early Regulations

At the same time that the demand for goods increased, the need for regulation arose. The first regulations were to control the quantities of goods that were being exchanged. Thus, in ancient Egypt government inspectors weighed and measured goods, as seen in Egyptian tomb paintings (Fig. 8-4). In ancient Greece and Rome, laws were enacted that governed the quality of wine and oil transported in *amphoras* (large ceramic jars). Measuring cups were certified by a "bureau of standards." In thirteenth century England, the size of bread loaves was regulated by law and bakers were required to stamp their marks into each loaf of bread.

In general, however, there was little official concern about product safety. The poor, the ignorant, and the gullible were constantly exploited. Milk and wine were watered down; chalk and clay were mixed with sugar and flour. Sellers of patent medicines offered concoctions that would supposedly cure all diseases known to humanity. At best, these potions contained harmless, if useless, ingredients; at worst, they contained dangerous drugs.

Early U.S. Legislation

An incident occurred during the Spanish-American War that became a major public scandal and marked a turning point in attitudes toward governmental responsibility for the safety of consumer products. Hundreds of soldiers died after eating canned beef that had been preserved in formaldehyde to speed up the sterilization process.

In the United States, Upton Sinclair's novel *The Jungle* exposed the unsanitary and dangerous methods used in slaughterhouses and meat-packing plants. Public reaction to the book was intense. In response, a group of reformers headed by Harvey Wiley, a chemist, and President Theodore Roosevelt drafted the Pure Food Law, which was passed by Congress in 1906. Later, the Standard Container Act of 1916 gave the U.S. Department of Agriculture control over the grading and labeling of all fresh and processed meat products. In time, other federal agencies, including the Environmental Protection Agency (EPA), Federal Trade Commission (FTC), and Food and Drug Administration (FDA), were formed to protect the consumer against fraud and hazardous merchandise.

It is important for designers to know about packaging laws and regulations and the federal agencies. No responsible manufacturer of consumer products will allow deceptive or unsafe packaging to be sold in the marketplace. A list of relevant sources of information is provided in the Appendix.

Current Laws and Regulations

In the United States today, every product is subject to both state and federal regulations governing packing, shipping, selling, advertising, labeling, testing, and marketing. Much of the legislation enacted in recent years has been intended to increase the protection of consumers against product-related hazards and fraud. Such legislation covers drugs, food, household products, toys, and cosmetics.

The Child Protection and Toy Safety Act of 1969, for example, bans toys and other children's articles that may pose electrical or mechanical hazards. The Fair Packaging and Labeling Act and the Hazardous Materials Control Act of 1970 address issues such as fireproofing textiles for children's clothing and producing safe toys, household products, and electrical appliances. Electrical appliances are tested by Underwriters Laboratories before being introduced to the public.

Issues related to trademarks, patents, copyrights, package and product safety, tampering, and child protection are subject to legal constraints and, therefore, require the guidance of an attorney. New copyright, trademark, and patent laws became effective on January 1, 1978.

Package Labeling

For the packaging designer, the most significant consumer protection legislation concerns labeling. Figures 8-5A and B illustrate government requirements for packaging, including labeling. It is absolutely crucial for the designer to observe these labeling laws and regulations in order to prepare correct comps and final

Some basic elements of correct labeling to meet FPLA requirements

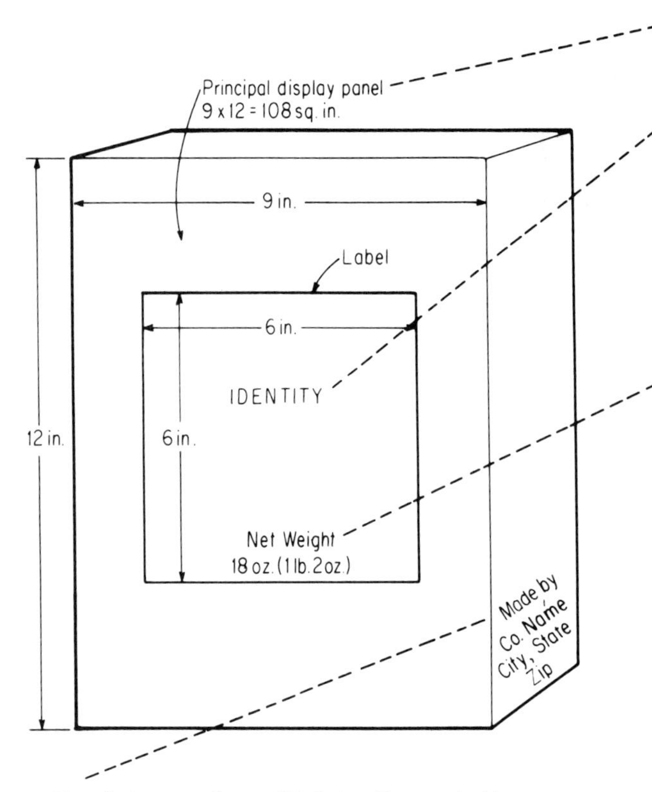

Panel most likely to be displayed, presented, shown or examined under normal and customary conditions of display. If more than one principal display panel is created, mandatory must be repeated.

Product name as required by applicable Federal law or regulation; or common or usual name; or either the generic name or a statement of function which appropriately describes the commodity. For drugs, a statement of the general pharmacological category or principal intended action should follow the "established name," common or usual name or proprietary name. The identity statement must be in lines generally parallel to the base on which the package or commodity rests as it is designed to be displayed. The identity statement must comprise a principal feature of the principal display panel and be in such type size and so positioned as to be easily read and understood.

Quantity statement can be left, right or centered; must appear somewhere within the lower 30% of the label. Must be parallel to the base of the package or commodity. Must be separated above and below from other printed matter by height of letters required and on each side by twice the width of the letter "N." Must be in a type size determined by area of side or surface of package or commodity that bears the principal display panel. Type size for this package (area of principal display panel = 108 sq. in.) must be at least ¼ in. for all characters including lower case. Dual declarations—total in ounces and largest whole unit—required for 1-lb. or more but less than 4-lb. packages: "NET WT. 18 OZ. (1 LB. 2 OZ.)" For liquids, 1 pint up to 1 gallon: "20 FL. OZ. (1 PT. 4 OZ.)" For lineal and square measurements, check regulations for specifics. Packages with principal display panel less than 5 sq. in. are exempt from lower 30% requirement, but remember, the area of the side of the package bearing the principal display panel is the determining factor for print size and separation.

Manufacturer, packer or distributor. Name and address may appear anywhere on the package. Packer's or distributor's name must be qualified, e.g., "Distributed by ________." Street may be omitted if address is in current city or telephone directory. City, state and zip code must be included. Only standard abbreviations permitted.

Four elements shown here illustrate basic requirements for correct packaging under FPLA. Foods, drugs and many other consumer products commonly found in supermarkets, drug, variety and other retail stores are involved. There are, of course, many variations depending on product and package and some exemptions, but this model will cover most basic situations. For questionable situations, consult FDA for foods, drugs, cosmetics and devices and FTC for all other consumer commodities.

Figure 8-5A. FTC regulations.

Determining type size of quantity statement to comply with Federal requirements (see table below)

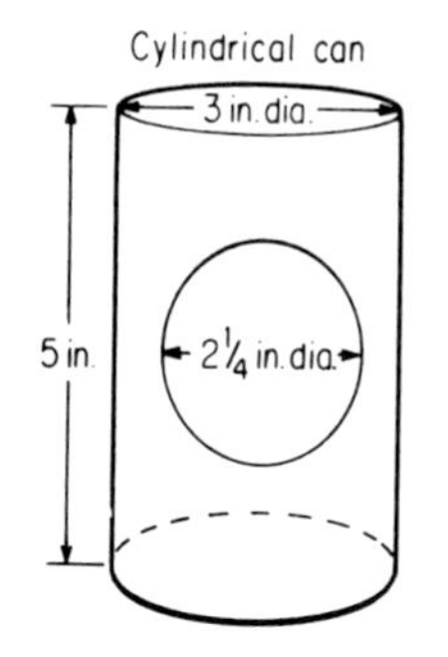

Principal display panel area is 40% of can height times circumference or 40% of 5 in. x 9.42 in. = 18.84 sq. in. (Circumference is 2 pi r, or 2 x 3.14 x 1.5.) Therefore, the required minimum type size to express quantity of contents on the label is ⅛ in., in spite of the label itself having an area of less than 5 sq. inches. When the expression takes the form of "Net Weight 12 ozs.," the lower case letters must meet the ⅛-in. requirement, making the upper case N and W somewhat higher than ⅛ inch.

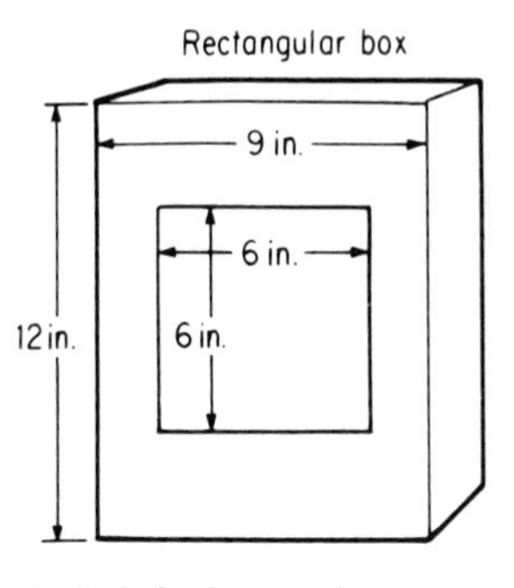

Principal display panel area is 9 in. x 12 in. or 108 sq. inches. Therefore, the minimum type size to express quantity of contents is ¼ inch. The size of the label (36 sq. in.) has no bearing on the minimum type size. The statement "NET WEIGHT 48 OZ. (3 LBS.)," in this instance, must have all letters and numerals ¼ in. high. If printed in upper and lower case, the lower case letters must be ¼ in., thus requiring the upper case N, W and O to be somewhat higher.

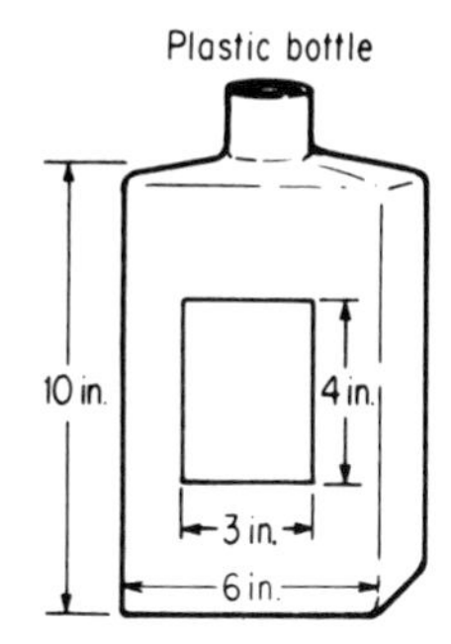

Principal display panel area is 10 in. x 6 in. = 60 sq. inches. (Tops, bottoms, flanges at the tops and bottoms of cans and shoulders and necks of bottles or jars are not considered part of the display panel.) A minimum type size of 3/16 in. is required to express quantity of contents on the label which itself only occupies 12 sq. inches. However, since regulations require the minimum type size of molded labeling to be increased by 1/16 in., the minimum type size on the molded flask must be ¼ inch.

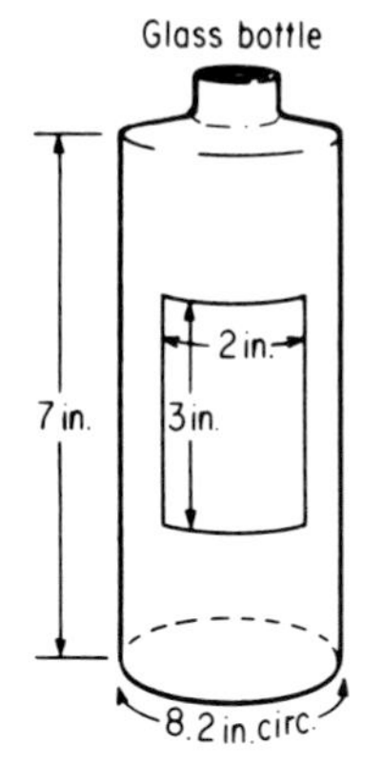

Principal display panel area is 40% of bottle height times circumference or 40% of 7 in. x 8.2 in. = 23 sq. inches. The minimum type size of the quantity of contents statement is ⅛ inch. The size of the label does not govern the minimum type size to be used for this container.

FPLA Requirements—minimum height of letters for quantity declaration

Square-inch area of main panel	Minimum height of numbers and letters	Minimum height: label information blown, formed or molded on surface of container
5 or less	1/16 in.	⅛ in.
More than 5, less than 25	⅛ in.	3/16 in.
More than 25, less than 100	3/16 in.	¼ in.
More than 100, less than 400	¼ in.	5/16 in.
More than 400	½ in.	9/16 in.

Figure 8-5B. FTC regulations.

art. Follow the instructions in Figures 8-5A and B to the letter.

Package Testing

Another topic of interest and concern to packaging designers is package testing. To ensure safe shipping and high-quality package materials, the packages and packaging materials must be tested in a laboratory before the package is manufactured. Several tests are applied, using specially designed equipment. The following are among the most frequently required tests.

Tensile and elongation test is basically a stretch test used for flexible packaging such as bags and films.

Impact test is used to measure the impact strength of films.

Tear test is a strength test for paper or film.

Stiffness test is performed with a Handle-O-Meter, which is used to determine the stiffness of packaging materials such as film and paper.

Water-vapor transmission test is for temperature, humidity, and moisture control.

Gas transmission test is used on films to determine their permeability to gases.

Bursting strength test is used for fiberboard, paperboard, corrugated cardboard, and bag stock.

Flat crush test is a procedure for testing corrugated board on a compression test device.

Fold endurance is a procedure to test double- and single-fold paper and board.

Grease penetration is a test for flexible materials intended to provide a barrier against greases.

Specular gloss is a test whereby a Glossmeter is used to specify and control surface characteristics of board and coated materials.

Conclusion

There is no excuse for ignorance of the law—so goes the common saying. That saying is as true for package designers as for those in other professions and for people in general. Therefore, this chapter has presented a brief overview of important laws and regulations regarding packaging material for designers. But keep in mind that with the ever increasing new packaging materials and systems, there will be new regulations by state and government agencies that will affect package design. To keep up with the laws, read the trade publications (listed on the next page), attend trade shows, and above all consult your supplier and packaging engineers.

REPORTS AND PUBLICATIONS

Foley, Patrick J. *Forecast of Plastic Materials Available Worldwide.* The Packaging Institute, USA, 1983.

Journal of Packaging Technology, Vol 2, No. 1, February 1988, p. 1.

Modern Plastics, *Guide to Plastic Packaging.* McGraw-Hill, 1988.

Refuse-Energy Systems with Resource Recovery as Alternate to Landfill, White, Weld & Co., New York, N.Y. 1977.

Numerous reports and publications are put out by the U.S. Department of Health, Education and Welfare: Bureau of Solid Waste Management and the U.S. Public Health Service, Environmental Health Service, Washington, D.C.

9

THE BEST OF THE BEST

A Showcase of Outstanding Packaging Design by Members of the Package Design Council International

WORKS by today's best package designers are presented on the following pages. They demonstrate the package designer's ability to further marketing and business goals and to do so with creativity and beauty.

With the goal of attracting the best young talent to the package design profession, the Package Design Council International (PDCI) sponsors conferences with educators on design training, helps schools expand their design curricula, and directs students toward the growing package design profession.

Fellowships in package design awarded by PDCI have enabled especially talented students to combine their graduate study with the opportunity to work with professional package designers. At meetings for young designers about to enter the profession, PDCI members review sample portfolios and discuss career opportunities in packaging design.

The Best of the Best is a publication of the Package Design Council International. Selections from the 1988 edition are reproduced here by permission of PDCI and the individual contributors.

Entrant:
Charles Biondo
Charles Biondo Design Associates, Inc.
New York, New York

Client:
Star Kist Foods
9 Lives

Objectives:
To redesign a line of canned cat food products which features "Morris," the brand's spokes-cat, for a closer association with the brand's semi-moist and dry product lines, which did not use the "Morris" image. "Morris" and a new logo for "9-Lives" are prominently featured on all packages, resulting in a uniform "Morris" presence in the sector.

Entrant:
Charles Biondo
Charles Biondo Design Associates, Inc.
New York, New York

Client:
General Foods Corporation
Impromptu

Objective:
Communicate product features: shelf-to-oven; ready-to-eat; entrees with a qualitative difference.

Entrant:
Charles Biondo
Charles Biondo Design Associates, Inc.
New York, New York

Client:
General Foods Corporation
Ronzoni

Objective:
To create a dominant brand synergy across two diverse product categories marketed in separate sectors: a line of over 100 dry pasta products and an extensive line of frozen entrees. To identify and differentiate products within the line. The "Pasta Semolina" seal was refined and moved to a central position to communicate the special product feature and enhance product identity.

Entrant:
Jack Schecterson
Herstein/Schecterson Inc.
New York, New York

Client:
Home Curtain Corporation
All Natural Cotton

Objective:
To introduce this new product called for strong visual POP impact and upscale imagery.

Entrant:
Jack Schecterson
Herstein/Schecterson Inc.
New York, New York

Client:
Boston Tea Company
Creme De Menthe Tea

Entrant:
Jack Schecterson
Herstein/Schecterson Inc.
New York, New York

Client:
Buddy L Corporation
Lil Brute Sets

Entrant:
Hans D. Flink
Hans Flink Design Inc.
White Plains, New York

Client:
Chesebrough-Pond's Inc.
Pond's Cosmetics Japan

Objective:
Competitive inroads prompted redesign of Pond's cosmetic line for the Japanese mass market. New upscale cartons feature containers shown in simple silhouette form against a decorative background with hot-stamped touches.

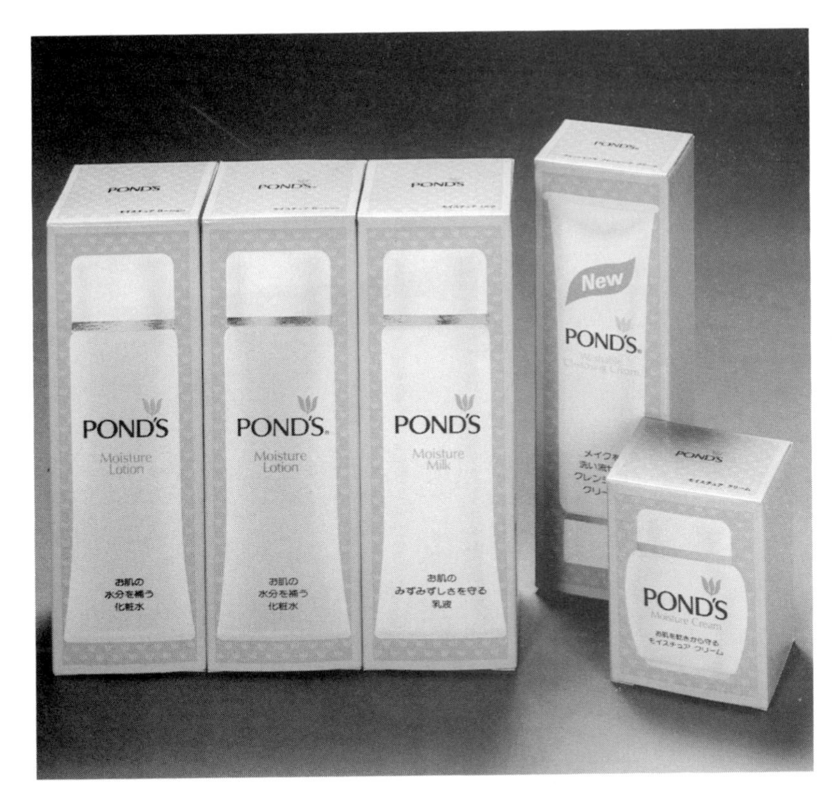

Entrant:
Hans D. Flink
Hans Flink Design Inc.
White Plains, New York

Client:
Richardson-Vicks Inc.
Tempo Soft Antacid

Objective:
This redesign features a new realistic and inviting product illustration and emphasized product claim copy. A much stronger logotype and brighter package coloration strengthen shelf impact.

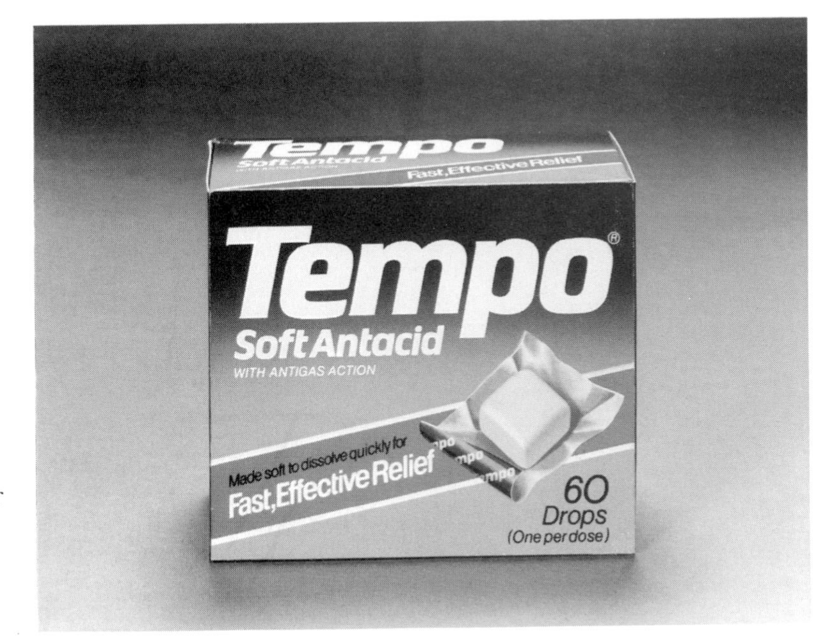

Entrant:
Edward Morrill
Coleman, Lipuma, Segal & Morrill, Inc.
New York, New York

Client:
The Gillette Co.
The Dry Look

Objective:
To present a unified contemporary, strongly masculine image for a line of hair care products for men. The redesigned format easily accepts additional new product forms and clearly differentiates between formulations.

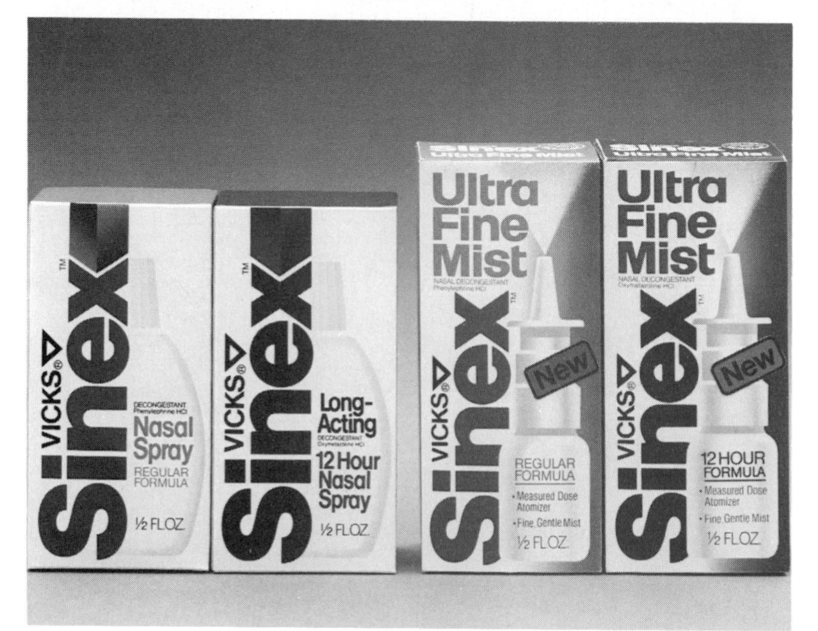

Entrant:
Hans D. Flink
Hans Flink Design Inc.
White Plains, New York

Client:
Richardson-Vicks Inc.
Vicks Sinex Product Line

Objective:
Packaging for the new nasal spray pump features product benefit "Ultra Fine Mist" by means of crisp airbrush illustration. Existing squeeze bottle packaging was redesigned into a vertical space saving format for a strengthened line identity and clarity of product differences.

Entrant:
Owen Coleman
Coleman, Lipuma, Segal & Morrill, Inc.
New York, New York

Client:
Thomas J. Lipton, Inc.
Wishbone Lite Salad Dressing

Objective:
A major design change across all lite Wishbone Salad Dressing labels to reflect an upscale contemporary "lite" image. The unique bulls-eye design utilizes color-coding for each flavor.

Entrant:
Owen Coleman
Coleman, Lipuma, Segal & Morrill, Inc.
New York, New York

Client:
The Quaker Oats Co.
Aunt Jemima Frozen Food Line

Objective:
Design development for the creation of a new upgrade umbrella image for Aunt Jemima Frozen Food Products. Initial conceptual work was executed on a graphics computer with the final design carried across 14 different products.

Entrant:
John Chrzanowski
Coleman, Lipuma, Segal & Morrill, Inc.
New York, New York

Client:
Konica U.S.A., Inc.
35mm Camera

Objective:
Konica required a strong, unified brand identity using the Konica logotype and "colorwheel" symbol together with an individual personality for each camera in its line of 35mm cameras. The design solution features transparent "blister-type" package structures which highlight the "designer style" cameras. The primary communication is the Konica brand name, complimented by impactful contemporary graphics and prioritized easy to read copy benefit points. All key communications in the purchasing decision aimed at the style conscious younger market, first time users "stepping-up" to 35mm cameras.

Entrant:
Barry G. Seelig
Apple Designsource, Inc.
New York, New York

Client:
Uncle Ben's Inc.
Uncle Ben's Rice—Boil-In-Bag

Objective:
Design a line extension that retains visual equity with the original product form while effectively communicating the difference of the line extension.

Entrant:
Stewart Mosberg
Dixon & Parcels
Associates Inc.
New York, New York

Client:
R. M. Palmer Company
Minty Bells

Objective:
The R.M. Palmer Company wanted to have an eye-catching and appetizing package which would be appropriate all seasons of the year. The marketing strategy suggested that the minty flavor of the chocolate had to be strongly communicated to the consumer. Dixon & Parcels chose a mint green bag with a clean window to show the colorfully wrapped "bells" to encourage impulse purchases.

Entrant:
J. Roy Parcels
Dixon & Parcels
Associates Inc.
New York, New York

Client:
Quaker State Inc.
Quaker State Oil Product Line

Objective:
To create a contemporary, clean, easily recognizable identification and image for all Quaker State products which would be equally effective for an integrated corporate identity system. Dixon & Parcels Associates refined the familiar "Q" and designed a special typeface for the words "Quaker State." Two gold racing stripes underline the logo relating the products to their use in automobile engines.

Entrant:
J. Roy Parcels
Dixon & Parcels
Associates Inc.
New York, New York

Client:
American Pop Corn Company
"American's Best" Jolly Time Pop Corn

Objective:
To effectively introduce "American's Best," a new high popping hybrid pop corn with a strong identification with the Jolly Time brand. An obvious distinction between the yellow and white popping corn had to be apparent. To communicate the quality and uniqueness while commanding attention and competitive distinctiveness, see-through plastic jars display the actual kernels while color coordinated caps and labels reinforce the difference of the two different types of popping corn.

Entrant:
Jack Schecterson
Herstein/Schecterson Inc.
New York, New York

Client:
Goldberger Doll
Manufacturing
Company, Inc.
Sandy Sandman

Objective:
Introducing a new licensed character doll.

Entrant:
David Pressler
Gerstman + Meyers Inc.
New York, New York
Designer: Larry Riddell

Client:
Kraft Dairy Group
Breyers Ice Cream

Objective:
Kraft's corporate goals included expanding its market for Breyers Ice Cream. Gerstman + Meyers' new packaging preserves equity and current users by revitalizing the Breyers leaf mark. Impactful black packages feature appetizing, larger-than-life scoops of ice cream and a new, sophisticated white Breyers logo to attract new customers across the U.S.A.

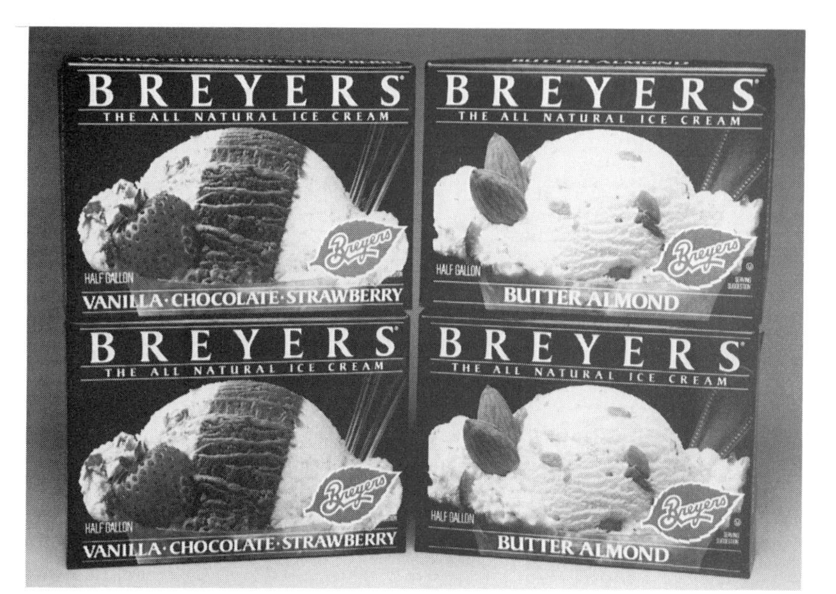

Entrant:
David Pressler
Gerstman + Meyers Inc.
New York, New York
Designer: Sabra Waxman

Client:
Plumrose
AM: Gourmet

Objective:
Gerstman + Meyers created unique packaging graphics for Plumrose's new AM: Gourmet line of frozen microwaveable breakfast entrees that establish a mini-section in the freezer case. A close-up photograph of a raw egg and eggshell, suggestive of a sunrise, combine with the digital-looking logo to evoke images of morning, wholesomeness, and breakfast.

Entrant:
David Pressler
Gerstman + Meyers Inc.
New York, New York
Designer: Lisa Lien

Client:
Luzianne Blue Plate Foods
Luzianne Cajun Creole Foods

Objective:
Luzianne Blue Plate Foods' new line of Cajun/Creole foods encompass a vast number of package configurations all tied together by a rich green background. Appetizing product photographs appear in the foreground with a romantic illustration of a southern plantation in the background at the top of the front panel. The new line now carves a strong position in a currently booming category.

Entrant:
David Pressler
Gerstman + Meyers Inc.
New York, New York
Designer: Sandy Meyers

Client:
Omni Hotels
Corporate identity program for the Omni Hotel chain

Objective:
Omni hotels management planned to expand its franchise to include a large number of hotels throughout the U.S. Gerstman + Meyers developed a corporate identity system based upon one universal logo and graphic symbol. An elegant mauve flower lock-up with each hotel name works well for individual hotels while providing a memorable image for the franchise.

Entrant:
Barry G. Seelig
Apple Designsource, Inc.
New York, New York

Client:
Madison's Inc.
Pasta Pour-over's

Objective:
The dominant black color on the label gives the product high shelf impact because of its distinctiveness and conveys a top quality image. Chef Tell's face appeals to many customers in the target audience who recognize him, and it helps identify the product with the person who developed it. The bottle shape conveys quality and uniqueness, and makes the bottle easier to handle.

Entrant:
David Scarlett; Bob Cruanas
Peterson & Blyth
Associates, Inc.
New York, New York

Client:
Flav-O-Rich Inc.
Flav-O-Rich

Objective:
Peterson & Blyth Associates created a brand identity for Flav-O-Rich, a line of ice cream from Dairymen, which repositions the products by communicating their high level of quality. The distinctive black packaging makes Flav-O-Rich stand out from the competition's predominately white packaging and the dramatic product photography gives the products a high level of appetite appeal.

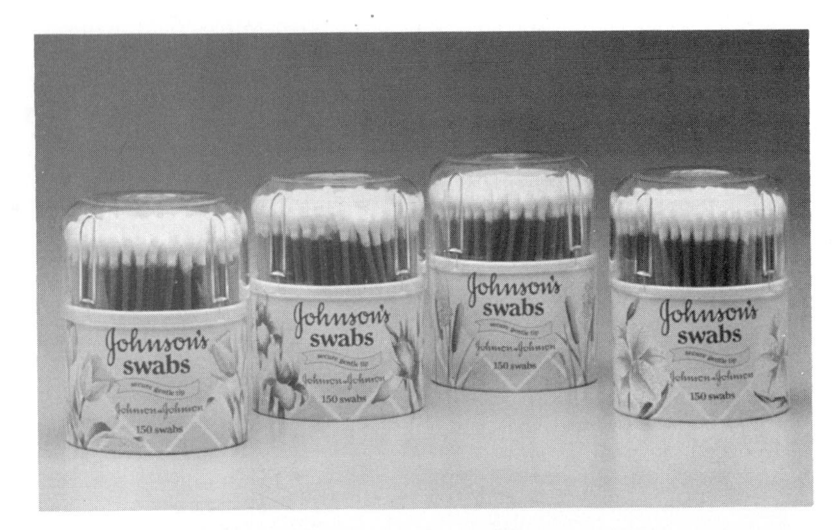

Entrant:
Barry G. Seelig
Apple Designsource, Inc.
New York, New York

Client:
Johnson & Johnson, Inc.
Johnson Swabs

Objective:
Design a line of decorative containers that gives the consumer a selection to fit within specific color schemes and personal tastes. The primary emphasis was to have the container placed within the usage area as a decorative and functional container.

Entrant:
Ronald Peterson; Bob Cruanas
Peterson & Blyth
Associates, Inc.
New York, New York

Client:
General Mills, Inc.
Clusters

Objective:
Peterson & Blyth Associates designed the bold, appetizing package design image that launched this new cereal from General Mills with a splash. Prominent, live action product photography heightens Clusters' appetite appeal, and the brand name—boldly set at the top of the package—gives this product introduction strong shelf impact in a highly competitive product category.

Entrant:
Barbara Wentz
Peterson & Blyth
Associates, Inc.
New York, New York

Client:
Hydron Ophthalmic Products
Hydron

Objective:
This line of contact lens care products projects a strong, efficacious image in a category where confusion reigns. The clean, ethical package design identity we created gives these products a trustworthy, well-established look that is highly appropriate for eye care products.

Entrant:
Barbara Wentz
Peterson & Blyth
Associates, Inc.
New York, New York

Client:
Personal Products Division,
3M Company
Buf-Puf Daily Cleanser

Objective:
The more upscale image we created for Buf-Puf Daily Cleanser communicates that this product line extension is not a sponge but rather a gentle method of deep cleansing the user's skin.

Entrant:
Richard Chesler/Ellie Rinner
The Design Source
New York, New York

Client:
Hygiene Industries
Body Brush

Objective:
To introduce the brush as part of a new product line while maintaining a strong visual alliance to the existing product line.

Entrant:
Richard Chesler/Ellie Rinner
The Design Source
New York, New York

Client:
Sony
Black Beauty

Objective:
Our challenge was to design a contemporary image for this family classic.

Entrant:
Richard Chesler/Ellie Rinner
The Design Source
New York, New York

Client:
New York Zoological Society
Project W.I.Z.E.—Survival Strategies

Objective:
To design and develop a modular system for zoo educational programs for 6th through 9th grade students to promote the idea of animal survival throughout the world.

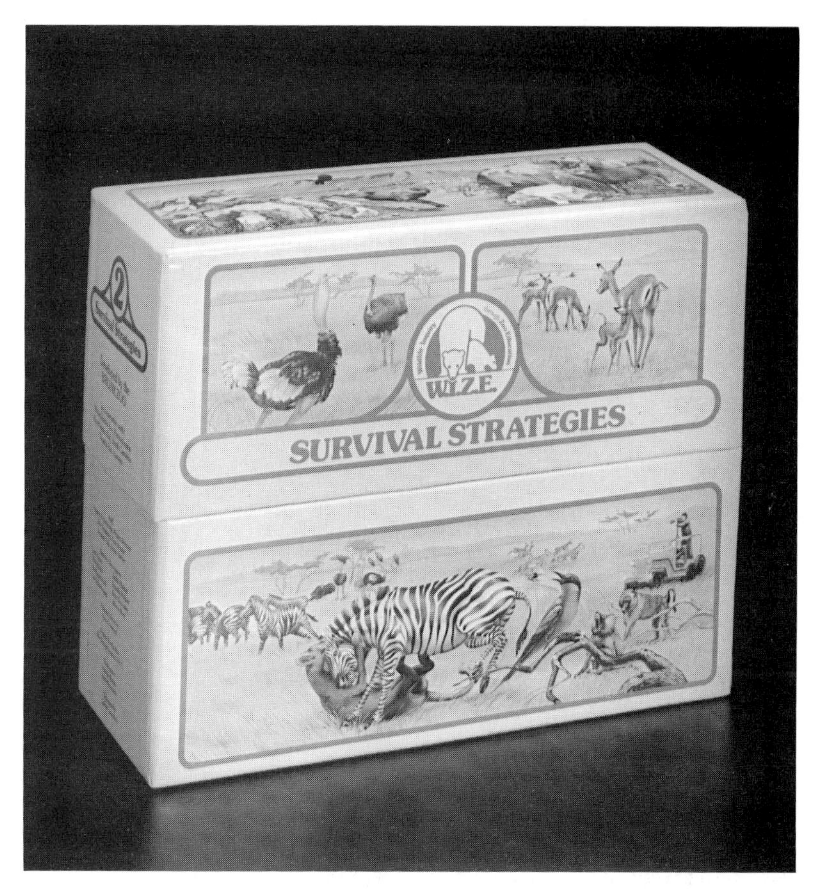

INDUSTRY ORGANIZATIONS

Adhesive and Sealant Council, Inc.
2350 East Devon Avenue
Des Plaines, IL 60018

American Paper Institute
260 Madison Avenue
New York, NY 10016

American Management Association
135 West 50th Street
New York, NY 10020

American Society for Testing Materials
1916 Race Street
Philadelphia, PA 19103

Association of Independent Corrugated Converters
2530 Crawford Avenue
Evanston, IL 60201

Can Manufacturers Institute
1625 Massachusetts Avenue NW
Washington, DC 20036

Cosmetic, Toiletry and Fragrance Association, Inc.
1625 Eye Street
Washington, DC 20006

Fiber Box Association, Inc.
224 South Michigan Avenue
Chicago, IL 60604

The Fragrance Foundation
116 East 19th Street
New York, NY 10003

Glass Packaging Institute
1800 K Street NW
Washington, DC 20006

National Association of Recycling Industries
360 Madison Avenue
New York, NY 10017

National Canners' Association
1133 20 Street NW
Washington, DC 20036

National Center for Resource Recovery
1211 Connecticut Avenue
Washington, DC 20036

National Flexible Packaging Association
12025 Shaker Boulevard
Cleveland, OH 44120

National Paper Box Association
231 Kings Highway East
Haddonfield, NJ 08033

National Soft Drink Association
1101 16th Street NW
Washington, DC 20036

National Solid Waste Management Association
1120 Connecticut Avenue NW
Washington, DC 20036

Package Design Council International
P.O. Box 3753
New York, NY 10017

Packaging Education Foundation
1700 K Street NW
Washington, DC 20006

The Packaging Institute USA
20 Summerset
Stamford, CN 06901

Packaging Machinery Manufacturers Institute
200 K Street NW
Washington, DC 20006

The Paper Bag Institute, Inc.
41 East 42nd Street
New York, NY 10017

Paperboard Packaging Council
1800 K Street NW
Washington, DC 20006

Point-of-Purchase Advertising Institute
60 East 42nd Street
New York, NY 10017

Society of Packaging and Handling Engineers
Reston International Center
Reston, Virginia 22091

Society of the Plastic Industry
250 Park Avenue
New York, NY 10017

FEDERAL AGENCIES

Consumer Product Safety Commission (CPSC)
1111 18th Street NW
Washington, DC 20207
(Hazardous household products, toys, hardware, flammable fabrics.)

Department of Agriculture (USDA)
14th Street and Independence Avenue SW
Washington, DC 20250
(Inspection of meat and poultry.)

Department of Labor, Occupational Safety and Health Administration (OSHA)
3rd Street and Constitution Avenue
Washington, DC 20210
(Packaging machinery and plant operations.)

Department of Transportation (DOT)
2100 Second Street SW
Washington, DC 20590
(Distribution of hazardous materials by land, sea, and air.)

Department of the Treasury
1200 Pennsylvania Avenue NW
Washington, DC 20226
(Imported goods, tobacco, alcohol, firearms. Sale and packaging regulations.)

Environmental Protection Agency (EPA)
401 M Street SW
Washington, DC 20460
(Pesticides, insecticides. Air, land, and water standards.)

Federal Trade Commission (FTC)
7th Street and Constitution Avenue NW
Washington, DC 20580
(Regulates deceptive packaging and labeling.)

Food and Drug Administration (FDA)
200 C Street SW
Washington, DC 20204
(Food, drugs, and cosmetics. Public health.)

National Center for Resource Recovery, Inc. (not a government agency)
1211 Connecticut Avenue
Washington, DC 20036
(Solid waste studies, management, and research. Recycling.)

Postal Service
475 L'Enfant Plaza SW
Washington, DC 20260
(Mailable goods. Postal regulations.)

APPENDIX **C**

PROFESSIONAL PUBLICATIONS

Aerosol Age
389 Passaic Avenue
Fairfield, NJ 07006

American Craft
401 Park Avenue South
New York, NY 10016

Boxboard Containers
300 West Adams Street
Chicago, IL 60606

Food and Drug Packaging
78500 Old Oak Boulevard
Middlesburg Heights, OH 44130

The Glass Industry
777 Third Avenue
New York, NY 10017

Industrial Design Magazine
330 West 42nd Street
New York, NY 10036

Modern Plastics
McGraw-Hill, Inc.
1221 Avenue of the Americas
New York, NY 10020

Package Printing
401 N. Broad Street
Philadelphia, PA 19108

Packaging
5 South Wabash Avenue
Chicago, IL 60603

Packaging Digest
400 N. Michigan Avenue
Chicago, IL 60611

Paperboard Packaging
747 Third Avenue
New York, NY 10017

Step-By-Step Graphics
6000 North Forest Park Drive
P.O. Box 1901
Peoria, IL 61656-9979

PACKAGING EDUCATION

California Institute of Art
24700 McBean Parkway
Valencia, CA 91255

Carnegie Melon University
Schenley Park
Pittsburgh, PA 15213

Clemson University
223 P&A Building
Clemson, SC 92631

Fashion Institute of Technology
227 West 27th Street
New York, NY 10001

Indiana State University
Terre Haute, IN 47809

Joint Military Packaging Training Center
Aberdeen Proving Grounds, MD 21005

Michigan State University
School of Packaging
East Lansing, MI 48824

New York University
80 Washington Square East
New York, NY 10003

Parsons School of Design
66 Fifth Avenue
New York, NY 10011

Pratt Institute
Graduate Design Programs
200 Willoughby Avenue
Brooklyn, NY 11205

Rhode Island School of Design
55 Canal Street
Providence, RI 02900

Rochester Institute of Technology
1 Lomb Memorial Drive
Rochester, NY 14623

Rutgers University
Packaging Science and Engineering
P.O. Box 909
Piscataway, NJ 08854

Sinclair Community College
Engineering, Packaging Technology
444 Third Street
Dayton, OH 45402

University of Wisconsin, Stout
Menominee, WI 54751

University of Missouri, Rolla
301 Harris
Rolla, MO 65401

University of New Haven
West Haven, CT 06516

GLOSSARY

Adhesion. Sticking objects together (as in pasting or gluing) with an adhesive material.

Aerosol. A pressurized container with a dispensing valve.

Ampoule. A small glass or plastic container used mainly for drugs and food ingredients. The ends are melted to seal the ampoule, and it is opened by breaking the stem.

Aseptic packaging. Sterilized containers made of plastic-lined paper-foil and plastic laminations. Aseptic packages have a long shelf-life and require no refrigeration in transport or in storage.

Blanks. Die-cut and scored paperboard ready to be assembled into cartons.

Blind embossing. *See* Embossing.

Blister pack. A transparent thermoformed shape attached to a card. There are several variations of this system.

Blow molding. A process of shaping plastics. Air is blown into a blob of molten plastic inside a mold. The plastic expands and takes the shape of the mold.

Calendering. Pressing thermoplastic material between two or more rolls to form a continuous sheet. Calenders are also used to coat papers and fabrics.

Caliper points. Units used to measure the thickness of paper. 0.0012″ equals 12 points.

Carboys. Large bottlelike containers made of glass or plastic usually encased in a wooden outer crate. They are used for shipping chemicals or other liquid products.

Cartons. Paper boxes basically of two types: the collapsible folding carton and the rigid, set-up paper box. Both types have several variations.

Catalyst. A hardener used to speed up polymerization.

Chipboard. Recycled paperboard. It is the lowest-cost board, and is adaptable for special linings.

Closures. Closing and sealing devices for bottles, jars, dispensers, and applicators.

Coextrusion. Two or more films are extruded simultaneously to form a multilayered film (*see* Extrusion).

Cohesion. Binding the surfaces of objects together through welding or use of a solvent and joining. When dry, the joint becomes a continuous piece of material.

Compression molding. A process of shaping plastics using heat and pressure.

Corrugated paperboard. A construction of alternate layers of flat and fluted paperboard. Flutes are categorized as A, B, C, or E.

Crimp. To squeeze or press the ends of tubes or cans using a series of folds or corrugations.

Die-cutting. Cutting shapes from paper, board, and plastics using cutting and stamping dies or lasers.

Drop test. A mechanical procedure to test the safety of a package's contents during shipping.

Electroplating. Coating plastic or other material with a thin layer of metal.

Embossing. A process of pressing paper between metal dies to create an image in relief on the paper. It can be used on either printed paper or blank paper (blind embossing).

Extrusion. A process of manufacturing rods, pipes, tubes, and film by feeding thermoplastic resin through a heated tube, then forcing the soft plastic through a die to form a continuous shape.

Films. Transparent or opaque flexible packaging materials.

Finish. The opening of a bottle designed to accommodate the closure.

Foils. Metallic-coated packaging papers.

Folding cartons. *See* Cartons.

Grain. The direction in which the fibers line up in paper.

Heatsealing. Process by which layers of plastic film are melted together to form a seal.

Hot stamping. Transferring a design from a thin foil or film onto a product through the application of heat and pressure.

Hydrocarbons. Chemical compounds used as aerosol propellants.

Injection molding. Molding method whereby melted resin is forced into the mold by a plunger.

In-mold process. Method by which a design is transferred to a plastic product during the molding process.

Labels. A variety of die-cut, self-adhesive applications to decorate or identify packages or products.

Laminating. Layers of material are impregnated with thermosetting resin, then pressed together, using a laminating press, to form a solid laminated mass.

Laser. A device that produces a very narrow beam of extremely intense light. Lasers are used in industrial processes and medicine and can also be used in packaging for die-cutting.

Master cartons. Larger sized cartons used in industry for shipping smaller cartons.

Multiwall bag. A flexible packaging form consisting of several layers of paper, plastic, or foil and used mainly for heavier bulk products.

Newsback. Chipboard with one side. It is used for inexpensive cartons, die-cuts, etc.

Paper bags. These come in several varieties, with or without side gussets (pleats). Paper bags are used for groceries, shopping bags, and inner containers.

Paperboard. Sometimes called *cardboard,* paperboard is made from laminated layers of paper in sheets of 0.0012″ (12 points) or more.

Parison. A thin-walled plastic tube placed in the mold. It is used in blow molding for making hollow shapes.

Plastisol molding. Coating or dip-molding nonplastic objects.

Polymers. Chemical compounds that make up plastics.

Pouch. Flexible container made from film, foil, or paper.

Private mold. A bottle mold created for an original concept.

Pyrolysis. A waste-recovery system that burns refuse in an oxygen-free atmosphere.

Recycling. The use of previously used materials to make new objects.

Resins. The raw materials for plastics.

Resource recovery. The use of materials that would otherwise go to waste.

Retort package. A polyester-foil-polypropylene laminate package with thermal adhesives. It allows the contents to be sterilized in the package.

Rotational molding. A slow molding process used for large objects. Plastic resins or liquids are placed in a hollow mold that is then rotated until all the material is fused to the mold's walls.

Shrink wrap or **shrink-film.** A low-cost

plastic wrap that is sealed around an object with heat.

Skin packaging. A method by which a thin plastic film is drawn over a product on a card.

Slush molding. Liquid plastisol is poured into a preheated hollow mold. After the material begins to gel, the excess liquid is dumped and the rest allowed to cool in the mold.

Stamping or **hot stamping.** A decorative process in which a roll of leaf is stamped with heated (or nonheated) metal dies.

Stock bottles. Bottles (glass or plastic) available ready-made in many shapes, sizes, and colors.

Thermoforming. The shaping of heated thermoplastic sheets or films through forced contact with the mold.

Thermoplastic polymers. Polymers that can be shaped and molded by heat. Since the bonds between molecules in these polymers are weak, they can be molded over and over again.

Thermosetting polymers. Polymers that set in a hard or rigid form. The bonds between molecules are strong, and these polymers do not soften when heated.

Transfer molding. A system of molding thermosetting materials. It is similar to compression molding.

Tubes. Extruded glass or plastic products used for packaging.

Tumblers. Glass or plastic containers shaped like drinking glasses with lids.

Typography. The style, arrangement, and appearance of typeset matter.

Universal product code (UPC). A code printed on packages that provides information on the product for inventory control and retail pricing.

Vacuum forming. *See* Thermoforming.

Vacuum metalizing. A process of coating plastics with metal.

Vial. A small glass container for medical and pharmaceutical products.

Welding. A method of heating or spinning the edges of two plastic containers or components and then joining them together.

INDEX